Biotechnology for Treatment of Wastes Containing Metals

RIVER PUBLISHERS SERIES IN CHEMICAL, ENVIRONMENTAL, AND ENERGY ENGINEERING

Series Editors

ALIREZA BAZARGAN
NVCo and University of Tehran
Iran

MEDANI P. BHANDARI
Akamai University, USA
Sumy State University, Ukraine
and Atlantic State Legal Foundation, NY, USA

HANNA SHVINDINA
Sumy State University, Ukraine

Indexing: All books published in this series are submitted to the Web of Science Book Citation Index (BkCI), to SCOPUS, to CrossRef and to Google Scholar for evaluation and indexing.

The "River Publishers Series in Chemical, Environmental, and Energy Engineering" is a series of comprehensive academic and professional books which focus on Environmental and Energy Engineering subjects. The series focuses on topics ranging from theory to policy and technology to applications.

Books published in the series include research monographs, edited volumes, handbooks and textbooks. The books provide professionals, researchers, educators, and advanced students in the field with an invaluable insight into the latest research and developments.

Topics covered in the series include, but are by no means restricted to the following:

- Energy and Energy Policy
- Chemical Engineering
- Water Management
- Sustainable Development
- Climate Change Mitigation
- Environmental Engineering
- Environmental System Monitoring and Analysis
- Sustainability: Greening the World Economy

For a list of other books in this series, visit www.riverpublishers.com

Biotechnology for Treatment of Wastes Containing Metals

Editor

Norma Gabriela Rojas Avelizapa

Instituto Politécnico Nacional
México

LONDON AND NEW YORK

Published 2019 by River Publishers
River Publishers
Alsbjergvej 10, 9260 Gistrup, Denmark
www.riverpublishers.com

Distributed exclusively by Routledge
4 Park Square, Milton Park, Abingdon, Oxon OX14 4RN
605 Third Avenue, New York, NY 10017, USA

First issued in paperback 2023

Biotechnology for Treatment of Wastes Containing Metals / by Norma
Gabriela Rojas Avelizapa.

Routledge is an imprint of the Taylor & Francis Group, an informa business

Publisher's Note
The publisher has gone to great lengths to ensure the quality of this reprint
but points out that some imperfections in the original copies may be
apparent.

While every effort is made to provide dependable information, the
publisher, authors, and editors cannot be held responsible for any errors
or omissions.

ISBN 13: 978-87-7022-954-8 (pbk)
ISBN 13: 978-87-7022-114-6 (hbk)
ISBN 13: 978-1-003-33738-6 (ebk)

Contents

Foreword

The treatment of water, sediments, soils or wastes containing metals represents a serious challenge mainly if the awareness for reducing or eliminating the generation of more pollution exists, as it is the case in some of the conventional treatments for residual wastes. In most cases, the pollutants or the risk associated with them will only transfer to other sites or media.

Since many decades ago, biotechnology has contributed as an eco-friendly, high efficiency alternative for minimizing, regenerating, or recycling residues, but biological processes face different and strong restrictions for the treatment of metal-containing materials.

For overcoming these restrictions, research in biotechnological alternatives continues and different approaches has been implemented since the isolation, modification, or adaptation of target microorganisms up to changes in media or the pre-treatment of residual wastes among many others. Additional alternatives include the use of metabolites or biomasses when the toxicity of media limits the microbial activity. Then, biotechnology can use different strategies or be utilized in combination with other less aggressive chemical or physical methods.

As it is well known, biological processes have been widely used for wastewater treatment, gas treatment, and the disposal of solid wastes in environmental engineering. Also, biological processes have been utilized for the production of biogas and hydrogen as new energy resources. In this book, the role of biotechnology on the treatment of residuals containing wastes is exposed.

The development and application of biotechnological methods for metal removal is essentially based on the use of the metal-microorganism mechanisms which have developed during their evolution and those related to metal-biomass interaction.

Biological treatment of metal-containing wastes is a multifactorial process since it depends on the composition of the residual materials, metal content and oxidation form, type of microorganism or by-product to be used; this is the main drawback of biotechnological processes since it is specific of the wastes to be treated.

The application of biotechnological technologies on metal-containing residual wastes has given excellent results in areas such as biomining, bioremediation, biosorption, bio-recovery. In some cases, could be possible to give an extra value to the residual wastes, situation that could be attractive for those companies that dispose high amounts of astes; recovered metals could be raw materials for other industrial processes.

There are many research opportunities in this area; this book aims to give an overview of the problems of metal pollution as well as the factors that restrict biological processes and information about how biotechnology could be applied or has been applied to reducing or eliminating metals contained in different kinds of residual wastes.

The diversity and perspectives of this expertise group in the treatment of metal-containing wastes make this book an important text for study and consultation to students, professors, and people who are concerned or involved in this topic.

Preface

This first edition of "Biotechnology for the treatment of wastes containing metals" addressed to teachers, researchers and students of training related to biotechnology applied to the treatment of materials containing metals is a book of general information and consultation for specialists and non-specialists. The book tries to update, deepen and extend these themes, taking into account the rapid evolution in this area of knowledge. This book tries to provide updated information with examples of applications in the bio treatment of wastes with a high content of metals. In this work, contributions have been gathered of researchers and specialists with different expertise related to the subject. In the chapters dedicated to basic knowledge, the focus is on updating information on the environmental problems and the impact that these wastes have on the environment, parameters to be considered previous the establishment of a biotechnological process also it is shown drawbacks and limitations for their success. Some other chapters deal with the use of biotechnology for the treatment of different types of matrices that contain metals, as well as mentioning the strategies that can be addressed to treat these materials effectively.

The collaborators belong to the different institutions, schools and faculties, for example, Friedrich Schiller University, Institut of Microbiology, Germany; Technological Institute of Durango; Technological University of the Metropolitan Area of the Valley of Mexico; Tecnológico de Monterrey, School of Engineering and Sciences; Institute of Biotechnology, Querétaro, Autonomous University of Nuevo Leon; Faculty of Chemical Sciences, Juarez University of the State of Durango; Faculty of Biological and Agricultural Sciences, México; Center for Research in Applied Science and Advanced Technology and the National School of Biological Sciences of the National Polytechnic Institute, México.

We hope that this book is a useful and accessible tool for teachers, students and professionals who are dedicated to the noble task of developing, implementing and applying biotechnology to the area of metal pollution.

The Authors

Acknowledgement

I would like to express my deep and sincere gratitude to Dr. Gómez-Ramírez, Dr. Rivas-Castillo, Dr. Medrano, Dr. Fierros-Romero, Dr. Kothe, Dr. Rojas-Avelizapa and MTA. Calvo and their research team for acceptance and contribute to the content of this book, providing invaluable time, information and data. It is great privilege and I deeply appreciate the trust received. I would also like to thank to Q.B.P. Olvera-Sandoval and Lic. Loa for support in formatting and adapting the content of the book.

List of Contributors

Andrea M. Rivas Castillo, *Universidad Tecnológica de la Zona Metropolitan del Valle de México, Boulevard, Miguel Hidalgo y Costilla No. 5, Los Héroes de Tizayuza, 43816, Tizayuca, Hidalgo, México;*
E-mail: a.rivas@utvam.edu.mx

Cuauhtémoc Contreras Mora, *Instituto Tecnológico Nacional de México-Instituto Tecnológico de Durango, Unidad de Posgrado, Investigación y Desarrollo Tecnológico (UPIDET), Avenida Felipe Pescador 1830 Ote. 34080 Durango, Durango, México; E-mail: cuauhtemoccon1@hotmail.com*

Damian Reyes Jáquez, *Instituto Tecnológico Nacional de México-Instituto Tecnológico de Durango, Unidad de Posgrado, Investigación y Desarrollo Tecnológico (UPIDET), Avenida Felipe Pescador 1830 Ote. 34080 Durango, Durango, México; E-mail: damian.reyes@itdurango.edu.mx*

David Enrique Zazueta Álvarez, *Instituto Tecnológico Nacional de México-Instituto Tecnológico de Durango, Unidad de Posgrado, Investigación y Desarrollo Tecnológico (UPIDET), Avenida Felipe Pescador 1830 Ote. 34080 Durango, Durango, México; E-mail: zazu_eta@hotmail.com*

Diana Alexandra Calvo Olvera, *Departamento de Biotecnología, Centro de Investigación en Ciencia Aplicada y Tecnología Avanzada del Instituto Politécnico Nacional, Cerro Blanco 141, Colinas del Cimatario, 76090, Santiago de Querétaro, Querétaro, México; E-mail: dcalvoo1500@alumno.ipn.mx*

Erika Kothe, *Friedrich Schiller University, Institute of Microbiology, Neugasse 25, D 07743 Jena, Germany; E-mail: erika.kothe@uni-jena.de*

Grisel Fierros Romero, *Tecnológico de Monterrey, School of Engineering and Sciences, Campus Querétaro, Avenida Epigmenio González No. 500, San Pablo, 76130, Querétaro, México; E-mail: gfierros@itesm.mx*

Hiram Medrano Roldán, *Instituto Tecnológico Nacional de México-Instituto Tecnológico de Durango, Unidad de Posgrado, Investigación y Desarrollo Tecnológico (UPIDET), Avenida Felipe Pescador 1830 Ote. 34080 Durango, Durango, México; E-mail: hiramdurango@yahoo.com.mx*

José Rubén Mundo Cabello, *Tecnológico de Monterrey, School of Engineering and Sciences, Campus Querétaro, Avenida Epigmenio González No. 500, San Pablo, 76130, Querétaro, México; E-mail: a01704376@itesm.mx*

Juan A. Rojas Contreras, *Department of Molecular Biology, Instituto Tecnológico de Durango, Boulevard Felipe Pescador 1830, Nueva Vizcaya 34080, Durango, Durango, México; E-mail: jrojas@itdurango.edu.mx*

Julia Kirtzel, *Friedrich Schiller University, Institute of Microbiology, Neugasse 25, D 07743 Jena, Germany; E-mail: julia.kirtzel@uni-jena.de*

Katrin Krause, *Friedrich Schiller University, Institute of Microbiology, Neugasse 25, D 07743 Jena, Germany; E-mail: katrin.krause@uni-jena.de*

Luis J. Galán Wong, *Instituto de Biotecnología, Universidad Autónoma de Nuevo León, Avenida Manuel L. Barragán, S/N Cd. Universitaria, San Nicolás de los Garza, N. L., México; E-mail: Javier.woong@gmail.com*

Luz Irene Rojas Avelizapa, *Facultad de Ciencias Biológicas y Agropecuarias-Córdoba, Universidad Veracruzana, Josefa Ortiz de Domínguez s/n. Peñuela, 94945, Amatlán de los Reyes, Veracruz; E-mail: luzrojas@uv.mx*

Marlenne Gómez Ramírez, *Departamento de Biotecnología, Centro de Investigación en Ciencia Aplicada y Tecnología Avanzada del Instituto Politécnico Nacional, Cerro Blanco 141, Colinas del Cimatario, 76090 Santiago de Querétaro, Querétaro, México; E-mail: mgomezr@ipn.mx*

Norma U. Estrada, *Facultad de Ciencias Químicas, Universidad Juárez del Estado de Durango, Avenida Veterinaria S/N Circuito Universitario, Valle del Sur 34120, Durango, Durango, México; E-mail: urtizn@hotmail.com*

Norma Gabriela Rojas Avelizapa, *Centro de Investigación en Ciencia Aplicada y Tecnología Avanzada del Instituto Politécnico Nacional, Colinas del Cimatario, 76090, Querétaro, Querétaro, México; E-mail: nrojasa@ipn.mx*

Reynaldo C. Pless, *Departamento de Biotecnología, Centro de Investigación en Ciencia Aplicada y Tecnología Avanzada del Instituto Politécnico Nacional, Cerro Blanco 141, Colinas del Cimatario, 76090 Santiago de Querétaro, Querétaro, México; E-mail: rcpless@yahoo.com*

Ricardo Serna Lagunes, *Facultad de Ciencias Biológicas y Agropecuarias-Córdoba, Universidad Veracruzana, Josefa Ortiz de Domínguez s/n. Peñuela, 94945, Amatlán de los Reyes, Veracruz; E-mail: rserna@uv.mx*

Sergio A. Tenorio-Sánchez, *Departamento de Microbiología, Escuela Nacional de Ciencias Biológicas, Instituto Politécnico Nacional, Prolongación de Carpio y Plan de Ayala s/n, Col. Santo Tomás, 11340, Ciudad de México, México; E-mail: stenorios@ipn.mx*

Susana Citlaly Gaucin Gutiérrez, *Instituto Tecnológico Nacional de México-Instituto Tecnológico de Durango, Unidad de Posgrado, Investigación y Desarrollo Tecnológico (UPIDET), Avenida Felipe Pescador 1830 Ote. 34080 Durango, Durango, México; E-mail: susana_gaucin@hotmail.com*

List of Figures

List of Tables

List of Abbreviations

AO	Overproduction of Antioxidants
CAZy	Carbohydrate-Active Enzymes Database
CDF	Cation Diffusion Facilitators
CFR	Code of Federal Regulations
CRT	Cathode Ray Tubes
DOD	U.S. Department of Defense
e.g.	*Exempli Gratia* in Latin
ECMM	Extracellular Mucilaginous Material
EGDE	Ethylene Glycol Diglycidyl Ether
EPA	Environmental Protection Agency
E-WASTE	*Electronic Waste*
FOLYMES	Fungal Oxidative Lignin Enzymes
HPC	Hydroprocessing Catalysts
i.e.	Id est in Latin
IARC	International Agency for Research on Cancer
ICP-MS	Inductively Coupled Plasma Mass Spectrometry
ICP-OES	Inductively Coupled Plasma Atomic Emission Spectroscopy
IOB	Iron Oxidizing Bacteria
LIBS	Lithium Ion Batteries
MICS	Minimum Inhibitory Concentrations
ORFS	Open Reading Frames
OSHA	Occupational Safety and Health Administration of the United States
PPR Ash	Power Plant Residual Ash
PTEs	Potentially toxic trace elements
PWBSS	Waste Printed Boards
RCRA	Resource Conservation and Recovery Act
Redox	Oxidation–Reduction
REE	Rare Earth Elements
ROS	Reactive Oxygen Species

SEM	Scanning Electron Microscope
SEM-EDX	Scanning Electron Microscope-Energy Dispersive X-Ray Analysis
SMRR	Specific Metal Removal Rate
SOB	Sulfur Oxidizing Bacteria
STM	Six Transmembrane Segments
UPIDET	Graduate Unit, Research and Technological Development
US	United States
VOCS	Volatile Organic Compounds
WHO	World Health Organization
XAFS	X-Ray Absorption Fine Structure
XANES	X-Ray Absorption Near Edge Structure
XRD	X-Ray Diffraction
YO	Oxygen Cell Yield

1

Overview of Metal Pollution

Marlenne Gómez Ramírez[1,*] and Sergio A. Tenorio Sánchez[2]

[1]Departamento de Biotecnología, Centro de Investigación en Ciencia Aplicada y Tecnología Avanzada del Instituto Politécnico Nacional, Cerro Blanco 141, Colinas del Cimatario, 76090 Santiago de Querétaro, Querétaro, México

[2]Departamento de Microbiología, Escuela Nacional de Ciencias Biológicas, Instituto Politécnico Nacional, Prolongación de Carpio y Plan de Ayala s/n, Col. Santo Tomás, 11340, Ciudad de México, México

E-mail: mgomezr@ipn.mx

*Corresponding Author

1.1 Introduction

Heavy metal is commonly defined as any metal and metalloid element that has a relative density ranging from 5.3 to 7 g/cm^3, atomic number above 20 and is toxic or poisonous at low concentrations (Abbas et al., 2014; Gautam et al., 2014). The main elements considered as heavy metals are As, Cr, Mn, Co, Cu, Zn, Mo, Hg, Ni, Sn, Pb, Cd, Sb, Tl, etc. There are three groups of heavy metals that are of concern, and they are the following: *toxic metals* (such as Hg, Cr, Pb, Zn, Cu, Ni, Cd, As, Co, Sn, etc.), *precious metals* (such as Pd, Pt, Ag, Au, Ru, etc.), and *radionuclides* such as U, Th, Ra, Am, etc. Four of them, Pb, Hg, Cd and Cr^{4+}, are known to have the highest toxicity among various metal ions, and the first three are in the limelight due to their major impact in the environment (Abbas et al., 2014).

Metals can be introduced into the human body through air, water, and food. There is enough evidence of how they play an essential role in the metabolism of living organisms at low concentrations but in higher concentrations can be toxic and represent serious health risk (Gautam et al., 2014). Although these metals have crucial biological functions in plants and animals, sometimes their chemical coordination and oxidation–reduction properties have given them an additional benefit so that they can escape control mechanisms such as homeostasis, transport, compartmentalization, and binding to required cell constituents. These metals bind with protein sites which are not made for them by displacing original metals from their natural binding sites causing malfunctioning of cells and, ultimately, toxicity. Previous research has found that oxidative deterioration of biological macromolecules is primarily due to binding of heavy metals to the DNA and nuclear proteins. Heavy metals are significant environmental pollutants and their toxicity is a problem of increasing significance for ecological, evolutionary, nutritional, and environmental reasons (Jaishankar et al., 2014).

The importance of metals and their toxic effect is becoming more relevant due to civilization, anthropogenic activities, and expansion of many industries (mining, metallurgical, electronic operation, surface finishing, energy and fuel producing, paper industries, fertilizer, pesticide, iron and steel, electroplating, electrolysis, etc.); a large variety of wastes are generated, which are dumped into the environment annually (Chauhan and Upadhyay, 2015; Fu and Wang, 2011; Gupta and Joia, 2016). From solid, liquid, and gaseous wastes, which are generated in majority (Sood and Chitre, 2017), the industrial wastes contain both valuable materials as well as hazardous materials which require special handling, dumping, and recycling methods (Chauhan and Upadhyay, 2015).

At present, the annual total solid waste generation worldwide is approximately 17 billion tons and it is expected to reach 27 billion by 2050 (Laurent et al., 2014). Currently, approximately 6×10^6 chemical compounds have been synthesized, and every year 1000 new chemicals are produced. About 60,000–95,000 of these compounds are in commercial use (Singh and Garima, 2014; Singh and Garima 2015), and this has been generating a considerable increase in the discharge of industrial wastes to the environment, mainly in soil and water, which has led to the accumulation of heavy metals in them. It is well known that these kinds of wastes, when applied as an improper disposal, become an important factor in metal contamination and cause serious environmental problems when they are leached into the environment (Jadhav and Hocheng, 2012).

The bad management of solid waste is an important contributor to many different environmental problems such as climate change (e.g., from emissions of greenhouse gases from landfills), stratospheric ozone depletion (e.g., from emissions of halocarbons in discarded cooling systems or in-use foams), human health damages (e.g., from exposure to chemicals and particles during waste collection and treatment), ecosystem damages (e.g., from emissions of heavy metals to air, soil and surface water), and resource depletion (e.g., due to currently inexistent or inefficient recycling systems for certain key minerals or metals) (Laurent et al., 2014). One of the main threats facing faced by the actual world (Girma, 2015a,b) is the contamination of groundwater, air, soil, and surface water due to the presence of heavy metals and contaminants which are transported by atmospheric particulates; this is an important pathway by which contaminants can be distributed in the environment. The transport of contaminants by air may occur by direct transfer of volatilized species or by particles. Atmospheric suspended particles (generally particle size of the range >60 μm), including aerosol and dust (referred to as atmospheric particulates), can play an important role in the transport of environmental contaminants, particularly those that have low volatility and low aqueous solubility and remain attached to soil particles (Csavina et al., 2012). On the other hand, plants require very small amounts of some metals for their growth and optimum performance, but their toxicity depends on various factors such as plant species, specific metal, concentration, chemical form, soil composition, and pH (Bhattacharjee and Goswami, 2018). Other metals are known to be toxic, such as Cr, Hg, Pb, Cd, Au, Ur, Zn, Se, Ni, Ag, and As. These heavy metals can cause the reduction of plant growth because they can affect the metabolic activities like plant mineral nutrition, photosynthesis, and activity of essential enzymes (Ojuederie and Babalola, 2017). But on the other hand, deficiencies of essential metals such Cu, Mn, Zn, and metalloids as Se in agricultural soils are affecting agricultural productivity. In addition to the above, the industrial revolution created an alarming situation for human life and aquatic biota because, since then, concentrations of several metals have increased in soil and water (Girma, 2015a,b).

Heavy metals are naturally present in the environment, soil, and food and are widely used in manufacturing processes as well as in the construction industry, causing them to be present in composted organic residuals (Smith, 2009). Some of the metals have extremely long biological half-life that essentially makes them cumulative toxics; also, some of them are carcinogenic in nature and can be transferred across the trophic levels of the food chain, reaching higher trophic levels causing them to bioaccumulate in living organisms (Fu and Wang, 2011; Gupta and Joia, 2016; Jadhav and Hocheng, 2012; Nancharaiah et al., 2016). Metals such as Cd, Pb, As, and Hg are strongly poisonous to metal sensitive enzymes, resulting in growth inhibition and death of organisms. The heavy metals such as Pb, Hg, and Ni are non-essential trace elements. These are highly toxic elements as they are persistent, bioaccumulative, and non-readily breakdown in the environment or not easily metabolized (Bhattacharjee and Goswami, 2018). There are many sources of heavy metals in products derived from household municipal solid wastes (MSW); these primarily can be classified into biodegradable (organic waste) and non-biodegradable (Plastic, glass, etc.). The main elements that generate concern include Zn, Cu, Ni, Cd, Pb, Cr, and Hg because they are potentially present (Smith, 2009; Sood and Chitre, 2017; Yuksel, 2015) and can come from batteries, waste mobile phones, sewage sludge, tannery waste, pig manure, poultry manure, disposable household materials, plastics, paints and inks, body care products, medicines, and household pesticides (Kim et al., 2018; Singh and Kalamdhad, 2012; Smith, 2009).

The major industrial wastes are produced from agricultural, electronic scraps, medical activity, metal finishing industry, industrial effluents, auto catalysts, manufacturing and recycling of batteries, fly ash, mining tailing, spent catalyst by petrochemical and petroleum refining industry. These solid wastes mostly contain Ag, As, Ba, Be, Cd, Co, Cu, Fe, Li, Mo, Mg, Zn, Cr, Hg, Ni, V, Pb, Se, Zn, Ti, etc. and precious metals such as Au, In, Ag, Pd, Pt, etc. (Arenas-Isaac et al., 2017a; Asghari et al., 2013; Chauhan and Upadhyay, 2015; Fornalczyk, 2012; Fu and Wang, 2011; Girma, 2015b; Gómez-Ramírez et al., 2014, 2015, 2018; Jadhav and Hocheng, 2012; Shailesh et al., 2016; Rivas-Castillo et al., 2018; Rojas-Avelizapa et al., 2018; Singh and Li, 2015). It is because of these waste materials that serious environmental problems are being generated, and these can be a potential source of heavy metals. This is the reason why these industrial wastes can be seen as a source of artificial minerals and valuable metals that can be recovered from them (Arenas-Isaac et al., 2017a; Fornalczyk, 2012; Girma, 2015b; Jadhav and Hocheng, 2012).

An appropriate approach to reduce the toxicity of these metals is the application of an adequate extraction process in order to recover some of the valuable metals present in them (Lee and Pandey, 2012). Different analytical methods have been applied for the characterization of solid wastes, from which X-rays have been successfully applied to characterize the chemical speciation of heavy metals. X-ray absorption fine structure (XAFS) spectroscopy, X-ray absorption near edge structure (XANES) spectroscopy, and extended X-ray absorption fine structure (EXAFS) spectroscopy can provide information about the distances, coordination number, and species of atoms surrounding the selected element, as well as the formal oxidation state and coordination chemistry (Zhang et al., 2013). According to the literature, the elemental composition of fresh, spent, and bioleached spent catalyst has been determined using different methods such as chemical analysis (ICP-OES), SEM photomicrograph, X-ray diffraction (XRD), scanning electron microscope-energy dispersive X-ray analysis (SEM-EDX), and CHN analysis after removing the organic content (Aung and Ting, 2005; Gholami et al., 2011; Mishra et al., 2007).

The analysis by ICP (inductively coupled plasma) is one of the most applicable analysis which has important role for determining the metal composition (Aung and Ting, 2005; Gómez-Ramírez et al., 2015, 2018a,b; Rojas-Avelizapa et al., 2013, 2018). During characterization of solid waste, it is important to know the associations with reactive phases and chemical forms (speciation) because depending on these is their release and environmental impact (Rodgers et al., 2015). Solid wastes as spent catalysts, e-waste, mining tailing, and landfill (complex mixture of organic and inorganic waste such as contaminated soil, oily sludge from car washes, organic materials, paper, wood, plastics, metals scrap, ash, medical waste, hazardous waste disposal, and wastes from municipalities, industries, and households) (Burlakovs et al., 2018) have different fractionation of metals and are divided into (i) exchangeable fraction, (ii) reducible fraction, (iii) oxidizable fraction, and (iv) residual fraction. In exchangeable fraction, the metals are highly mobile and easily affected by the ionic composition of water, which are acid-soluble and/or bound to carbonate. Reducible fraction involves metals bound to Fe-Mn oxide and are susceptible to leaching in an acidic environment.

The exchangeable and reducible fractions are related to weekly adsorbed elements retained on the surface by relatively weak electrostatic attraction. The oxidizable fraction contains metals that are bound to the organic matter and/or sulfide and require highly oxidizing conditions for their liberation from the matrix. The residual fraction is the most stable form of the metals and comprises those that are incorporated into the crystal lattice of minerals and do not release them under normal environmental conditions. The metals in the residual fraction are strongly bound, whereas metals in the other fractions bind weakly through electrostatic attraction (Burlakovs et al., 2018; Ferreira et al., 2017; Pathak et al., 2018; Srichandan et al., 2014). It is important to know the content of metals in the different fractions because the water-soluble and exchangeable metal forms are generally more toxic than other forms because they can be easily released into water. Metal ions can inactivate enzymes by reacting with sulfhydryl groups which are an integral part of the catalytically active sites of these biomolecules or are involved in maintaining the correct structural relationship of the enzyme protein. Metals can also reduce enzyme activity by interacting with the enzyme–substrate complex by denaturing the enzyme protein or by interacting with the protein active groups. Thus, the activity of microorganisms would be lower as the levels of water-soluble and exchangeable metal become higher (Tripathy et al., 2014). In the case of coal combustion wastes as ashes and slags, the chemical composition of slags and ashes strongly depends on the type of material that is burned and the incineration technology and has different fractions such as exchangeable fraction, fraction bound to carbonates, fraction bound to iron and manganese oxides, fraction bound to organic matter, and residual fraction (Karwowska et al., 2015).

Due to previous exposure, different methods such as physical, chemical, biological, or hybrid methods have been used for treatment of solid wastes containing metals. It is very important that schemes of solid wastes handling improve in implementing advanced systems for recovery and reuse of various materials. Nowadays, the "zero waste" concept is becoming more topical through the reduction of disposed waste. Recovery of metals, nutrients, and other materials that can be returned to the material cycles still remains as a challenge for future (Burlakovs et al., 2018). The Basel convention was adopted internationally during the 1990s as a worldwide response to the need to control the disposal of waste movements between countries, regulate the destination of waste, and where necessary, its movement and transportation (Ziglio, 2014). It is important the existence of environmental legislation an the the documentation of those polluted areas. The scarce information will no allow to control the exposure which will bring complications in the future due to the negative effects of heave metals. Occupational exposure to heavy metals can be decreased by engineering solutions.

Monitoring the exposure and probable intervention for reducing additional exposure to heavy metals in the environment and in humans can become a momentous step towards prevention. National as well as international cooperation is vital for framing appropriate tactics to prevent heavy metal toxicity (Jaishankar et al., 2014).

1.2 Conclusion

Metals participate in different metabolic processes in living organisms, how-ever, in high concentrations or by their nature can be toxic to them. There are metals known as heavy metals which cause damage to the cells affecting their activity and development and therefore to the organisms. These type of metals can be found in different types of waste and can be released by mishandling them, becoming a risk to the environment. That is why it is necessary to establish policies as well as more appropriate processes for the treatment of wastes containing these metals and to prevent them from reaching the environment and thus not affecting animals, humans, plants, and other living organisms.

References

Abbas, S.H., Ismail, I.M., Mostafa, T.M., Sulaymon, A.H., 2014. Biosorption of heavy metals: A review. J. Chem. Sci. Technol. 3, 30.

Arenas-Isaac, G., Gómez-Ramírez, M., Montero-Álvarez, L.A., Tobón-Avilés, A., Fierros Romero, G., Rojas-Avelizapa, N.G., 2017. Novel microorganisms for the treatment of Ni and V as spent catalysts. Indian J. Biotechnol. 16, 370–379.

Asghari, I., Mousavi, S.M., Amiri, F., Tavassoli, S., 2013. Bioleaching of spent refinery catalysts: A review. J. Ind. Eng. Chem. 19, 1069–1081. https://doi.org/10.1016/j.jiec.2012.12.005

Aung, K.M.M., Ting, Y.-P., 2005. Bioleaching of spent fluid catalytic cracking catalyst using *Aspergillus niger*. J. Biotechnol. 116, 159–170. https://doi.org/10.1016/j.jbiotec.2004.10.008

Bhattacharjee, T.M., Goswami, M.M., 2018. Heavy metals (As, Cd & Pb) toxicity & detection of these metals in ground water sample. A review on different techniques. Int. J. Eng. Sci. Invention, 7(1), 12–21.

Burlakovs, J., Jani, Y., Kriipsalu, M., Vincevica-Gaile, Z., Kaczala, F., Celma, G., Ozola, R., Rozina, L., Rudovica, V., Hogland, M., Viksna, A., Pehme, K.-M., Hogland, W., Klavins, M., 2018. On the way to 'zero waste' management: Recovery potential of elements, including rare earth elements, from fine fraction of waste. J. Clean. Prod. 186, 81–90. https://doi.org/10.1016/j.jclepro.2018.03.102

Chauhan, R., Upadhyay, K., 2015. Removal of heavy metal from E-Waste: A review. Int. J. Chem. Stud. 3, 15–21.

Csavina, J., Field, J., Taylor, M.P., Gao, S., Landázuri, A., Betterton, E.A., Sáez, A.E., 2012. A review on the importance of metals and metalloids in atmospheric dust and aerosol from mining operations. Sci. Total Environ. 433, 58–73. https://doi.org/10.1016/j.scitotenv.2012.06.013

Ferreira, P.F., Sérvulo, E.F.C., Da Costa, A.C.A., Ferreira, D.M., Godoy, M.L.D.P., Oliveira, F.J.S., 2017. Bioleaching of metals from a spent diesel hydrodesulfurization catalyst employing *Acidithiobacillus thiooxidans* FG-01. Braz. J. Chem. Eng. 34, 119–129. https://doi.org/10.1590/0104-6632.20170341s20150208

Fornalczyk, A., 2012. Industrial catalysts as a source of valuable metals. J. Achiev. Mater. Manuf. Eng. 55, 6.

Fu, F., Wang, Q., 2011. Removal of heavy metal ions from wastewaters: A review. J. Environ. Manage. 92, 407–418. https://doi.org/10.1016/j.jenvman.2010.11.011

Gautam, R.K., Sharma, S.K., Mahiya, S., Chattopadhyaya, M.C., 2014. Contamination of heavy metals in aquatic media: Transport, toxicity and technologies for remediation. Heavy Metals in Water: Presence, Removal and Safety. pp. 1–24. https://doi.org/10.1039/9781782620174-00001

Gholami, R.M., Borghei, S.M., Mousavi, S.M., 2011. Bacterial leaching of a spent Mo–Co–Ni refinery catalyst using *Acidithiobacillus ferrooxidans* and *Acidithiobacillus thiooxidans*. Hydrometallurgy 106, 26–31. https://doi.org/10.1016/j.hydromet.2010.11.011

Girma, G., 2015a. Microbial bioremediation of some heavy metals in soils: An updated review. J. Resour. Dev. Manag. 10, 62-73–73.

Girma, G., 2015b. Microbial bioremediation of some heavy metals in soils: An updated review. Egypt. Acad. J. Biol. Sci. G Microbiol. 7, 29–45. https://doi.org/10.21608/eajbsg.2015.16483

Gómez-Ramírez, M., Rivas-Castillo, A.M., Monroy-Oropeza, S.G., Escorcia-Gómez, A., Rojas-Avelizapa, N.G., 2018a. Effect of glucose concentration on Ni and V removal from a spent catalyst by *Bacillus* spp. strains isolated from mining sites. Acta Univ. 28, 1–8. doi:10.15174/ au.2018.1475

Gómez-Ramírez, M, Rivas-Castillo, A., Rodríguez-Pozos, I., Avalos-Zuñiga, R.A., Rojas-Avelizapa, N.G., 2018b. Feasibility study of mine tailing's treatment by *Acidithiobacillus thiooxidans* DSM 26636. Int. J. Biotechnol. Bioeng. 12, 4. https://doi.org/10.5281/zenodo.2363155

Gómez-Ramírez, M., Montero-Álvarez, L.A., Tobón-Avilés, A., Fierros-Romero, G., Rojas-Avelizapa, N.G., 2015. *Microbacterium oxydans* and *Microbacterium liquefaciens*: A biological alternative for the treatment of Ni-V-containing wastes. J. Environ. Sci. Health Part A Tox. Hazard. Subst. Environ. Eng. 50, 602–610. https://doi.org/10.1080/10934529.2015.994953

Gómez-Ramírez, M., Zarco-Tovar, K., Aburto, J., de León, R.G., Rojas-Avelizapa, N.G., 2014. Microbial treatment of sulfur-contaminated industrial wastes. J. Environ. Sci. Health Part A Tox. Hazard. Subst. Environ. Eng. 49, 228–232. https://doi.org/10.1080/10934529.2013.838926

Gupta, A., Joia, J., 2016a. Microbes as potential tool for remediation of heavy metals: A review. J. Microb. Biochem. Technol. 8. https://doi.org/10.4172/1948-5948.1000310

Gupta, A., Joia, J., 2016b. Microbes as potential tool for remediation of heavy metals: A review. J. Microb. Biochem. Technol. 8. https://doi.org/10.4172/1948-5948.1000310

Jadhav, U.U., Hocheng, H., 2012. A review of recovery of metals from industrial waste. J. Achiev. Mater. Manuf. Eng. 54, 9.

Jaishankar, M., Tseten, T., Anbalagan, N., Mathew, B.B., Beeregowda, K.N., 2014. Toxicity, mechanism and health effects of some heavy metals. Interdiscip. Toxicol. 7, 60–72. https://doi.org/10.2478/intox-2014-0009

Karwowska, E., Wojtkowska, M., Andrzejewska, D., 2015. The influence of metal speciation in combustion waste on the efficiency of Cu, Pb, Zn, Cd, Ni and Cr bioleaching in a mixed culture of sulfur-oxidizing and biosurfactant-producing bacteria. J. Hazard. Mater. 299, 35–41. https://doi.org/10.1016/j.jhazmat.2015.06.006

Kim, Y., Seo, H., Roh, Y., 2018. Metal recovery from the mobile phone waste by chemical and biological treatments. Minerals 8, 8. https://doi.org/10.3390/min8010008

Laurent, A., Bakas, I., Clavreul, J., Bernstad, A., Niero, M., Gentil, E., Hauschild, M.Z., Christensen, T.H., 2014. Review of LCA studies of solid waste management systems – Part I: Lessons learned and perspectives. Waste Manag. 34, 573–588. https://doi.org/10.1016/j.wasman.2013.10.045

Lee, J., Pandey, B.D., 2012. Bio-processing of solid wastes and secondary resources for metal extraction – A review. Waste Manag. 32, 3–18. https://doi.org/10.1016/j.wasman.2011.08.010

Mishra, D., Kim, D.J., Ralph, D.E., Ahn, J.G., Rhee, Y.H., 2007. Bioleaching of vanadium rich spent refinery catalysts using sulfur oxidizing lithotrophs. Hydrometallurgy 88, 202–209.

Nancharaiah, Y.V., Mohan, S.V., Lens, P.N.L., 2016. Biological and bioelectrochemical recovery of critical and scarce metals. Trends Biotechnol. 34, 137–155. https://doi.org/10.1016/j.tibtech.2015.11.003

Ojuederie, O.B., Babalola, O.O., 2017. Microbial and plant-assisted bioremediation of heavy metal polluted environments: A review. Int. J. Environ. Res. Public. Health 14. https://doi.org/10.3390/ijerph14121504

Pathak, A., Healy, M.G., Morrison, L., 2018. Changes in the fractionation profile of Al, Ni, and Mo during bioleaching of spent hydroprocessing catalysts with *Acidithiobacillus ferrooxidans*. J. Environ. Sci. Health Part A Tox. Hazard. Subst. Environ. Eng. 53, 1006–1014. https://doi.org/10.1080/10934529.2018.1471033

Rivas-Castillo, A.M., Gómez-Ramirez, M., Rodríguez-Pozos, I., Rojas-Avelizapa, N.G., 2018. Bioleaching of metals contained in spent catalysts by *Acidithiobacillus thiooxidans* DSM 26636. Int. J. Biotechnol. Bioeng. 12, 5. https://doi.org/10.5281/zenodo.2021685

Rodgers, K.J., Hursthouse, A., Cuthbert, S., 2015. The potential of sequential extraction in the characterisation and management of wastes from steel processing: A prospective review. Int. J. Environ. Res. Public. Health 12, 11724–11755. https://doi.org/10.3390/ijerph 120911724

Rojas-Avelizapa, N.G., Hipólito-Juárez, I.V., Gómez-Ramírez, M., 2018. Biological treatment of coal combustion wastes by *Acidithiobacillus thiooxidans* DSM 26636. Mex. J. Biotechnol. 3(3), 54–67.

Rojas-Avelizapa, N.G., Gómez-Ramírez, M., Hernández-Gama, R., Aburto, J., García de León, R., 2013. Isolation and selection of sulfur-oxidizing bacteria for the treatment of sulfur-containing hazardous wastes. Chem. Biochem. Eng. Q. 27, 109–117.

Shailesh, D, Monal, S., Devayani, T., 2016. E-Waste: Metal pollution threat or metal resource? J. Adv. Res. Biotechnol. 1, 1–14. https://doi.org/10.15226/2475-4714/1/2/00103

Singh, J., Kalamdhad, A.S., 2012. Reduction of heavy metals during composting – A review. Int. J. Environ. Prot. 12(9), 36–43.

Singh, N., Li, J.H., 2015. Bio-extraction of metals as secondary resources from E-Waste [WWW Document]. Appl. Mech. Mater. 768, 602–611. https://doi.org/10.4028/www.scientific.net/AMM.768.602

Singh S.P., Garima, T., 2014. Application of bioremediation on solid waste management: A review. J. Bioremed. Biodeg. 05. https://doi.org/10.4172/2155-6199.1000248

Smith, S.R., 2009. A critical review of the bioavailability and impacts of heavy metals in municipal solid waste composts compared to sewage sludge. Environ. Int. 35, 142–156. https://doi.org/10.1016/j.envint.2008.06.009

Sood, G., Chitre, N., 2017. Review of biological methods for hazardous waste treatment. IOSR J. Biotechnol. Biochem. 6, 50–55. https://doi.org/10.9790/264X-06025055

Srichandan, H., Pathak, A., Kim, D.J., Lee, S.-W., 2014. Optimization of two-step bioleaching of spent petroleum refinery catalyst by *Acidithiobacillus thiooxidans* using response surface methodology. J. Environ. Sci. Health Part A Tox. Hazard. Subst. Environ. Eng. 49, 1740–1753. https://doi.org/10.1080/10934529.2014.951264

Tripathy, S., Bhattacharyya, P., Mohapatra, R., Som, A., Chowdhury, D., 2014. Influence of different fractions of heavy metals on microbial eco-physiological indicators and enzyme activities in century old municipal solid waste amended soil. Ecol. Eng. 70, 25–34. https://doi.org/10.1016/j.ecoleng.2014.04.013

Yuksel, O., 2015. Influence of municipal solid waste compost application on heavy metal content in soil. Environ. Monit. Assess. 187, 313. https://doi.org/10.1007/s10661-015-4562-y

Zhang, H., Yao, Q., Zhu, Y., Fan, S., He, P., 2013. Review of source identification methodologies for heavy metals in solid waste. Chin. Sci. Bull. 58, 162–168. https://doi.org/10.1007/s11434-012-5531-2

Ziglio, L., 2014. Industrial solid waste management in Brazil and the basel convention. Novos Estud. Jurídicos 19, 585. https://doi.org/10.14210/nej.v19n2.p585-606.

2

Environmental Impacts of Solid Waste Containing Metals

Norma U. Estrada[1], Andrea M. Rivas Castillo[2,*] and Juan A. Rojas Contreras[3]

[1]Facultad de Ciencias Químicas, Universidad Juárez del Estado de Durango, Avenida Veterinaria S/N Circuito Universitario, Valle del Sur 34120, Durango, Durango, México
[2]Universidad Tecnológica de la Zona Metropolitan del Valle de México, Boulevard, Miguel Hidalgo y Costilla No. 5, Los Héroes de Tizayuza, 43816, Tizayuca, Hidalgo, México
[3]Department of Molecular Biology, Instituto Tecnológico de Durango, Boulevard Felipe Pescador 1830, Nueva Vizcaya 34080, Durango, Durango, México
E-mail: a.rivas@utvam.edu.mx
*Corresponding Author

2.1 Introduction

One of the main challenges faced by both developing and already developed countries is the protection of the environment and human health assurance. The main purpose of regulating waste elimination is to reduce the introduction of polluting substances that can represent risk factors for ecosystems' health. However, current environmental policies and legislation are, for the most part, focused on regulating waste disposal instead of addressing and preventing its generation (EPA, 2019).

The Resource Conservation and Recovery Act (RCRA), stated in 1976, created the framework for hazardous and non-hazardous waste management programs in the United States (U.S). The materials regulated by RCRA are known as "solid waste". Only materials that meet the RCRA definition of solid waste can be classified as hazardous waste, which are subjected to additional regulations. The Environmental Protection Agency (EPA) developed detailed regulations to define what materials qualify as solid wastes and hazardous wastes. Understanding the definition of a solid waste is an important first step in the process that EPA has set for generators to track hazardous wastes by determining whether the waste generated is a regulated hazardous waste. The RCRA states that "solid waste" means any waste or sludge from a wastewater treatment plant, water supply treatment plant, or air pollution control facility and other discarded material that results from industrial, commercial, mining or agricultural operations, and from community occupations. It is important to keep in mind that the definition of solid waste is not limited to physically solid waste. Many wastes are liquid, semi-solid, or contain gaseous material. Thus, a solid waste is any material that is discarded because it is:

(a) Abandoned: the term abandoned means thrown away. A material is abandoned if it is discarded, burned, incinerated, or recycled in a simulated manner.

(b) Inherently similar to waste: some materials pose such a threat to human health and the environment that they are always considered solid waste; these materials are considered similar to wastes, such as wastes those containing dioxins.

(c) Discarded military ammunition: military ammunition encompasses all the products and components of ammunition produced or used by the U.S. Department of Defense (DOD) or the U.S. Armed Services for defense and national security.

(d) Recycled in certain ways: a material is recycled if it is used or reused (i.e., as an ingredient in a process), reclaimed, or used in certain ways (it is used on the land in a manner that constitutes a waste, it is burned to recover energy, or it is accumulated).

Exclusions specific to the definition of solid waste are listed in the Code of Federal Regulations (CFR). Many of these exclusions are related to recycling. Materials that do not meet this definition are not solid waste and are not subjected to the RCRA regulation (EPA, 2018a).

2.2 Environmental Impacts of Mining Waste

Mining and metallic mineral processing are one of the main environmental pollution causes. The metallic mineral extraction through conventional mining procedures involves large quantities of rock extraction from ground. Open-pit mining (less than 100 m from the surface) produces greater amounts of overburden than underground mining. In mining areas, the environmental problem is mainly related to mechanical landscape and the acid drainage produced by the mines. Approximately 60% of disturbed areas are used for overburden elimination, which represent approximately 40% of the annual solid waste generated in the U.S. Although mines are classified according to their predominant product, they can produce large quantities of other elements as co-products. As a result, the processing of metal ores generally leads to multi-elemental environmental pollution (Lua and Caib, 2012). In general, the economically important minerals occur as metal sulfides, whose exploitation causes SO_2 release into the atmosphere and acid accumulation near foundry operations. Smelting continues to be a major source of gaseous and dust pollutants. Copper smelting is considered as one of the main SO_2 contamination sources caused by humans. Consequently, soil near foundries have low pH values (Terraglio and Manganelli, 1967), and water pH also decreases in lakes close to exploitation areas (Chon and Hwang, 2000).

Generally, smelter emissions and acidic soil conditions cause a reduction in the abundance and diversity of soil microorganisms. This reduction decreases soil fertility by interrupting the biogeochemical cycles of C and N (Fernandes et al., 2018). Primary metal smelters greatly contribute to vegetation damage due to SO_2 contamination, soil acidification, and metal contamination (Terraglio and Manganelli, 1967). The uptake of trace elements from contaminated soils and the direct deposition of pollutants from the atmosphere on the plants' surface can lead to their contamination. As a consequence, the plants accumulate toxic levels of trace elements and become contaminant transfer factors along the food chain. Arid soils due to vegetation loss are especially susceptible to erosion, which leads to a greater environmental damage in the vicinity of foundries (Hailemariam et al., 2017).

Practically, all metals are used for commercial purposes. The main metals produced in large quantities are Fe, Al, Cu, Pb, Mn, and Zn. The remaining elements that are extracted and recovered in smaller quantities are called minor metals. Among the main metals, the production of Cu, Pb, and Zn causes the greatest environmental concern. In the group of minor metals, As and Cd are the most dangerous due to their geochemical and toxicological properties (Singh et al., 2011). Relevant metals often found as pollutants due to mining activities are described in the following.

2.2.1 Copper (Cu)

Cu occurs naturally in a wide variety of minerals, of which chalcopyrite (Cu sulfide) is the most common. The largest Cu deposits are found in the

U.S, Chile, Canada, the Commonwealth of Independent States, Zambia, and Peru (Cox et al., 1981), and most of these deposits are extracted in open wells, generating every day huge amounts of wastes from mines and landfills, more than the exploitation of any other metal (EPA, 2018b). Landfills produced by the exploitation of Cu sulfide represent up to 98% of the volume of ores and are eliminated in the form of large piles or valley fillings (Chikusa, 1994). Cu open-cast mines are rarely filled after mining because they are so large and ore bodies are not near the surface. Modern mines are reforested and recovered when they are abandoned, as landfills are. This does not happen to old Cu mines, which are a rich source of metals, acid water, and sediments. The environmental concern for the production of Cu has focused on the SO_2 emissions and the easily vaporized trace elements (As, Cd, Hg, etc.) from the smelters. However, Cu emissions are also a serious problem since more than 65% of this kind of anthropogenic emissions come from the smelters (Dudka and Adriano, 1997).

2.2.2 Lead (Pb) and Zinc (Zn)

Pb and Zn are often found together in mineral deposits, but they have different applications and biological effects. Zn is a physiologically essential element, while Pb does not have a known positive biological function and creates serious environmental and health hazards. As a result, Zn primary mining production has been increasing over the past decade, while Pb production has begun to decline (Dudka and Adriano, 1997). Pb smelter and refinery emissions are tightly regulated, as are the workers' blood Pb levels when in regular contact with this element (Karrari, 2012). Although this latter practice elevates the cost of Pb production, it is necessary to protect both the general public health and workers' health in the Pb industry. Pb is obtained from the galena and Zn comes from the sphalerite (Sclzaefer et al., 2004). Pb and Zn are extracted in 47 countries and in 50 countries, respectively, which makes them the most widespread metals in terms of primary production (Dudka and Adriano, 1997). The secondary Pb smelters used to process Pb metallic products are found in 43 countries, but only 21 countries have secondary Zn smelters, reflecting the large recycling of Pb in batteries for electrical storage (WHO, 2017). The extensive recycling of Pb results in the production of smelting metal, which is substantially higher than mining production. Pb and Zn processing is an important historical source of environmental contamination (Dudka and Adriano, 1997). Metal-rich dispersion zones consisting of galena particles around the smelter or mine contain relatively immobile Pb, while the Pb oxides and sulfates of the combustion gases are much more soluble and can be moved by rainwater to deeper levels of the soil. They also disperse further from the source due to their small particle size. Although smelter emissions have decreased significantly in the U.S in recent years, most of these emissions have not reached the limits of 0.15 and 1.5 g/m^3 for Pb (EPA, 2016). The most important Pb market is the one of electric power batteries, which currently represents 80% of the U.S Pb consumption. On the other hand, Zn is used to make galvanized steel and various Zn alloys (Dudka and Adriano, 1997).

2.2.3 Arsenic (As)

As is well known for its toxicity (WHO, 2018). Large doses of As (>100 mg per person) induce acute As poisoning, resulting in death (Ratnaike, 2003). As is mainly used in wood preservatives, herbicides, and insecticides, and it is largely recovered as a by-product along with other metals due to its appearance (Fowler et al., 2007). This metal vaporizes at 615°C and, therefore, is released during roasting of base metal ores. Older smelters do not capture this As vaporization in scrubbers and are surrounded by As-rich zones. Around the Cu-Ni smelters, As concentration in lake sediments deposited over the past decades is several times higher than that in the times before the foundry began (EPA, 1976).

2.2.4 Cadmium (Cd)

The use of Cd, Ni, Au, or Hg rechargeable batteries in power electric cars can help to solve the fossil fuels' crisis, but Cd may be potentially considered as one of the most toxic trace elements in the environment (Chrzan, 2016). Cd is particularly dangerous due to its easy absorption by plants and its tendency to accumulate in crops that are part of the food chain. Besides, Cu has a persistent nature once it is in the environment. Despite environmental concerns, Cd production has not decreased and, therefore, 60% of the total Cd input into the air comes from smelting and mining activities. The metal is recovered from the combustion dust during the roasting and sintering of sphalerite and in the sludge from Zn electrolytic refining. With the production of 1 mg of Zn, 3 kg of Cd are produced (Schmidt, 2002). Therefore, the environmental problems caused by Zn production are, at least in part, attributed to Cd emissions to the environment.

2.3 Potential Environmental Contaminants Associated with Electronic Waste

As a result of the rapid development of information technology and the constant updating of electronic products, waste known as E-waste has become the fastest growing municipal solid waste stream in the last decade (Schmidt, 2002; Ni et al., 2009; Robinson, 2009). E-waste is chemically and physically different from other forms of municipal or industrial waste since it contains both valuable materials and hazardous materials that require special handling and special recycling methods in order to avoid environmental contamination and their harmful effects on human health. E-waste includes waste of electronic products such as televisions and cell phones. Besides, the rapid growth of computing has driven the production of this type of waste, as it is estimated that more than one billion computers have been removed in recent years. The majority of electronic waste is produced by developed countries, in which only the U.S generates more than 9.4 million tons (LaDou and Lovegrove, 2008). However, due to the lack of facilities for electronic waste management, high labor costs, and strict environmental regulations, developed countries tend not to recycle electronic waste. Instead, these wastes are currently deposited in landfills or exported (around 60%) to developing countries (e.g., China, India, and Vietnam), where they can be recycled using inadequate techniques and without any or little consideration for workers' health care and for the protection of the environment (Cobbing, 2008). Although according to the Basel Convention of 1992 the unregulated export of electronic waste is illegal, it continues through clandestine operations and legal loopholes (UNEP, 2009). The environmental effects of electronic waste disposal are attracting increasing attention from politicians, non-governmental organizations, and the scientific community as well because, since 2006, there have been more than 500 scientific articles related to E-waste and its environmental effects.

The chemical composition of the electronic waste varies according to the type of the discarded article. Most electronic waste consists of a mixture of metals, particularly Cu, Al, and Fe; bonded, covered, or mixed with various types of plastics and ceramics (Hoffmann, 1992). A discarded personal computer with a weight of 25 kg consists of metal (43.7%), plastics (23.3%), electronic components (17.3%), and glass (15%). However, laptops can contain a higher concentration of heavy metals (Berkhout and Hertin, 2004). Virtually, all electronic waste contains some valuable components or base metals. These wastes may provide an incentive for recycling in poor countries, otherwise, they could represent an elevated risk for human health or environmental deterioration. Metals such as Cu and Pt are included among the components of electrical contacts due to their high electricity conductance and chemical stability. The concentration of precious metals in the circuit boards of printers is more than 10 times higher than that of the minerals extracted commercially (Betts, 2008). The concentrations of Cd, Cu, Ni, Pb, and Zn are such that, if these elements were released into the environment, they would pose a risk to ecosystems and human health (Wilmoth et al., 1991). Although some contaminants can be eliminated by recycling, large quantities of these residues end up concentrated in landfills or electronic waste recycling centers, where they can still negatively affect human health or the environment. It is believed that electronic waste is an important contributor to the large amounts of Cu emitted annually into the environment (Bertram et al., 2002).

China is one of the largest E-waste recyclers in the world and receives more than one million tons of E-waste from the U.S and European countries each year. Despite the economic benefits, electronic waste has caused serious environmental contamination in China as it is often illegally processed in patios or workshops using primitive techniques such as open burning, strong acid digestion or dismantled into components to recover precious metals (Zhang and Min, 2009). Worse still, these workshops are usually located close to arable land, affecting crops and, therefore, human health through the food chain (Luo et al., 2011; Zheng et al., 2013). In view of the increasing domestic generation of E-waste and the demand for high-quality electronic products (Yang et al., 2008), E-waste problems in China will be growing. As such, the Chinese government has modified the laws, regulations, and standards to reduce the number of illegal E-waste recycling workshops (Lu et al., 2015).

2.4 Waste from Refinery Spent Catalysts

Hydroprocessing catalysts (HPC) are widely used in the oil refining industry, mainly to purify hydrocarbon processing streams. These solid catalysts play an important role since they are mainly used to improve the efficiency of the process, to remove sulfur from the stream, and to break the long-chain hydrocarbons to form sulfur-free short-chain hydrocarbons (Ancheyta and Speight, 2007). There are several types of catalysts that are used, but the main ones are hydroprocessing catalysts, fluid catalytic cracking catalysts, and reforming catalysts. During the feeding of crude oil, the catalysts are contaminated with impurities and get deactivated. For example, diesel hydro-desulfurization catalysts generally have a life cycle of 3–4 years (Chiranjeevi et al., 2016).

Worn-out HPC are produced in petrochemical industries and classified as hazardous solid wastes according to EPA due to the presence of hazardous materials that include heavy metals (like V and Ni), metal oxides, and metal sulfides. These catalysts have very high porosity and surface areas, in which the coke is deposited and the catalyst is deactivated. Once the catalysts complete their life cycle, these catalysts will be removed from the process and, at this stage, the catalysts are considered as "spent" (EPA, 2003).

The use of HPC has increased drastically due to strict regulations on mandatory sulfur content (sulfur-free or ultra-low sulfur content) in petroleum products such as diesel and jet fuels (Dufresne, 2007). For example, the maximum sulfur content in ultra-low diesel fuel is 10–15 ppm (Rashid and Sayed, 2011). According to the literature, it is estimated that the total amount of spent hydroprocessing catalysts generated worldwide is in the range of 150,000–170,000 tons per year. Therefore, an increase in the production and use of HPC is anticipated for at least the next 10 years, and it is estimated that the demand for these catalysts in the market will grow at an annual rate of 5%. The exact quantities of spent catalysts are not available, but considering the short life of the catalyst and the approximately 400 units operating worldwide, the volumes are assumed to be very high (Dufresne, 2007). The market for fresh hydrotreating catalysts is currently around 120,000 tons per year. It has been stated that 50% of this amount is used to clean the fuels produced as column distillates and the remaining 50% is used to update and purify the waste. The same trend is expected for the market of hydrocracking catalysts, which currently hovers around 10,000 tons per year and is expected to grow at a rate of around 5% per year (Furimsky, 2007).

The increasing rate in the use of fresh catalyst is the most significant factor for the increment of spent catalysts; however, there are many other factors on which the total volume of HPC spent that is discarded as solid waste depends. These parameters include, but are not limited to:

- Increased need for new catalysts in hydroprocessing units to meet the growing demand for production of ultra-low sulfur fuels.
- Reduction of cycle times due to severe operations in diesel hydrotreating units to comply with strict fuel specifications.
- A greater demand in the processing of crude with sulfide high content and metals based on economic criteria and heavier feeding currents.

- Reduced catalyst deactivation times due to a faster deactivation rate and the lack of availability of reactivation processes for HPC waste.

The quantities of spent catalysts discarded from the hydroprocessing units are always higher than the amounts of fresh catalyst used in that unit due to the formation of coke and metal deposits on the surface of these catalysts (Ferguson, 1990). These metal complexes and the metals present in fresh catalysts are potential hazards and are subjected to EPA environmental regulations.

2.5 Effects of Heavy Metals on the Environment and Health

In recent years, the ecological concern and of the public health has grown due to the presence of contamination by heavy metals. The term "heavy metal" refers to any metallic chemical element that has a relatively high density compared to water and can be toxic or poisonous to living beings (Duffus, 2002). Within the group of heavy metals, metalloids are also included. That is the case of As, which can induce toxicity at low exposure levels (Bradl, 2002).

Human exposure to heavy metals has increased dramatically as a result of an exponential increase in their use in various industrial, agricultural, domestic, and technological applications (Bradl, 2002). Heavy metals are natural elements found on the Earth's surface. However, most metallic environmental pollution and human exposure result from anthropogenic activities, such as mining and smelting, industrial production, and domestic and agricultural use of metals and metal-containing compounds (Shallari et al., 1998; Herawati et al., 2000; Goyer, 2001; He et al., 2005).

Some metals, like Co, Cu, Cr, Fe, Mg, Mn, Mo, Ni, Se, and Zn are considered essential micronutrients required for various biochemical and physiological processes. Inadequate supply of these elements results in a variety of deficiencies, diseases, or syndromes (WHO, 1996). Other metals such as Al, Sb, As, Ba, Be, Bi, Cd, Ga, Ge, Au, In, Pb, Li, Hg, Ni, Pt, Ag, Sr, Te, Tl, Sn, Ti, V, and U do not have established biological functions and are not considered essential metals (Nordberg et al., 2015).

Due to their high degree of toxicity, metals such as As, Cd, Cr, Pb, and Hg are among the priority metals of great importance for public health. All of them are systemic toxins that are known to induce multiple organ damage, even at low exposure levels. However, it is known that each one has unique characteristics and physicochemical properties that give them specific toxicological mechanisms of action. The following is a summary of their environmental occurrence, production, uses, and potential risk to human health (Singh et al., 2011).

2.5.1 Arsenic (As)

As is an element that is detected in low concentrations in practically all environmental matrices. The environmental pollution caused by As is due to natural phenomena such as volcanic eruptions, soil erosion, and anthropogenic activities (ATSDR, 2000). As is used to make products with agricultural applications such as insecticides, herbicides, fungicides, etc. Some As compounds have also been used in the medical field for the treatment of syphilis, amoebic dysentery, and trypanosomiasis (Tchounwou et al., 1999; Centeno et al., 2006).

It is estimated that millions of people are chronically exposed to As worldwide, especially in developing countries, such as Bangladesh, India, Chile, Uruguay, and Mexico, where groundwater is contaminated with high concentrations of As. Exposure to As occurs orally (ingestion), inhalation, dermal contact, and parenteral route (NRCC, 1978), and its concentrations on air vary from 1 to 3 ng/m^3 in remote locations (far from human emissions) and from 20 to 100 ng/m^3 in cities. As concentration in water is usually less than 10 µg/L, although higher levels can occur near natural mineral deposits or mining sites. Its concentration in diverse foods varies from 20 to 140 ng/kg (Morton and Dunnette, 1994). For most individuals, diet is the main source of exposure, with an average consumption of approximately 50 µg per day. Intake by air, water, and soil is usually much smaller, but exposure to these media can become significant in areas of recognized As contamination.

Pollution with high levels of As is worrisome because As can cause a series of negative effects on human health. Several epidemiological studies have reported a strong association between As exposure and increased carcinogenic risks and systemic health effects (Tchounwon et al., 2003b). Interest in As toxicity has been increasing by recent reports from large populations in Bangladesh, Thailand, Inner Mongolia, Taiwan, China, Mexico, Argentina, Chile, Finland, and Hungary that have been exposed to high concentrations of As in water and are showing various clinical and pathological conditions, including cardiovascular and peripheral vascular diseases, developmental anomalies, neurological and neurobehavioral disorders, diabetes, hearing loss, portal fibrosis, hematological disorders (anemia, leukopenia, and eosinophilia), and carcinoma (NRC, 2001; Tchounwon et al., 2003).

Research has also reported significantly higher standardized mortality rates for bladder, kidney, skin, and liver cancers in many As contaminated areas.

2.5.2 Cadmium (Cd)

Cd is a heavy metal of considerable environmental and occupational concern. It is widely distributed on the Earth's crust at an average concentration of approximately 0.1 mg/kg. The highest level of Cd compounds in the environment accumulates in sedimentary rocks (GESAMP, 1987).

Cd is frequently used in various industrial activities. The main industrial applications of Cd include the production of alloys, pigments, and batteries (Tchounwou et al., 2012). While the use of Cd in batteries has shown considerable growth in recent years, its commercial use has declined in developed countries in response to environmental concerns. In the U.S, for example, the daily Cd intake is approximately 0.4 µg/kg/day, less than the half of the oral reference dose of EPA. This decrease has been linked to the introduction of general restrictions on Cd consumption (EPA, 2000).

The main routes of Cd exposure are inhalation or cigarette smoke and food ingestion, while absorption through the skin is rare. Human exposure to Cd is possible through a variety of sources, including employment in the primary metals' industries, consumption of contaminated food, working in workplaces contaminated with Cd, and cigarette smoking, which, as mentioned before, is the most important contributor (Paschal et al., 2000; IARC, 2018). Other Cd sources include emissions from industrial activities, like mining, smelting, batteries' manufacturing, pigments, stabilizers, and alloys (ATSDR, 2012a). Cd is also present in trace amounts in certain foods such as leafy vegetables, potatoes, grains and seeds, liver and kidney, and crustaceans and molluscs (Satarug et al., 2003). In addition, foods that are rich in Cd can greatly increase the concentration of this element in the human body. Examples are liver, mushrooms, shellfish, mussels, cocoa powder, and dried seaweed.

Cd chronic exposure by particles' inhalation is generally associated with changes in lung function and chest radiographs consistent with emphysema (Davison et al., 1988). Occupational exposure to Cd particles in air has been associated with decreases in olfactory function (Mascagni et al., 2003). Several epidemiological studies have documented an association of chronic low-level Cd exposure with decreased bone mineral density and osteoporosis (Akesson et al., 2006; Gallagher et al., 2008; Schutte et al., 2008).

Exposure to Cd is commonly determined by measuring Cd levels in blood or urine. Blood Cd reflects recent exposure to Cd (from smoking, for example). Cd in urine (usually adjusted by dilution when calculating the Cd/creatinine ratio) indicates accumulation or renal Cd loading (Järup, 1998; Wittman and Hu, 2002). It is estimated that around 2.3% of the U.S population has high levels of Cd in urine (>2 μg/g creatinine), a marker of chronic exposure and body burden (Becker et al., 2002). Thus, Cd levels in blood and urine are typically higher in cigarette smokers, intermediate in former smokers, and lower in non-smokers (Becker et al., 2002; Mannino et al., 2004). Due to the continued use of Cd in industrial applications, Cd environmental contamination and human exposure have increased dramatically (Elinder and Järup, 1996).

Cd is a serious lung and gastrointestinal irritant, which can be fatal if inhaled or swallowed. Generally, after 15–30 min of an acute Cd ingestion, there appear symptoms such as abdominal pain, burning sensation, nausea, vomiting, salivation, muscle cramps, dizziness, shock, loss of consciousness, and seizures. Acute Cd ingestion can also cause gastrointestinal tract wear, lung, liver or kidney damage, or even coma, depending on the poisoning route (Baselt, 2017). Studies in rodents have shown that Cd chronic inhalation causes pulmonary adenocarcinomas (Waalkes, 2000). After systemic administration or direct exposure, it can also cause proliferative prostate lesions that include adenocarcinomas (Waalkes and Rehm, 1992).

Several regulatory bodies classify Cd compounds as human carcinogens. The International Agency for Research on Cancer (IARC) and the U.S. National Toxicology Program have concluded that there is adequate evidence to sustain that Cd is a human carcinogen. This designation as a human carcinogen is mainly based on repeated findings of an association between Cd occupational exposure and lung cancer as well as experiments with rodents that show that the lung system is a Cd objective site (Waalkes, 2000).

2.5.3 Chrome (Cr)

Cr is a natural element present on the Earth's surface, with oxidation states (or valence states) ranging from Cr (II) to Cr (VI) (Jacobs and Testa, 2005). Cr compounds are stable in the [Cr (III)] trivalent form and appear naturally in this state in minerals, such as ferrochromiume The hexavalent form [Cr (VI)] is the second most stable state (Waalkes, 2000). On the other hand, elemental Cr [Cr (0)] does not occur naturally. Cr enters several environmental matrices (air, water, and soil) from a wide variety of natural species and anthropogenic sources, with greater release from industrial establishments. The industries with the greatest contribution of Cr release include metal processing, tannery facilities, chromate production, stainless steel welding, and the production of ferrochrome and Cr pigment. The increase in the environmental Cr concentrations has been related to Cr release in air and wastewater, mainly from the metallurgical, refractory, and chemical industries. This Cr liberation from anthropogenic activities into the environment occurs mainly in the hexavalent form [Cr (VI)] (ATSDR, 2012a). Hexavalent Cr [Cr (VI)] is a toxic industrial pollutant classified as a human carcinogen by several regulatory and non-regulatory organisms (IARC, 1990; ATSDR, 2012a). The health hazard associated with Cr exposure depends on its oxidation state, ranging from the low toxicity of the metallic form to the high toxicity of the hexavalent form. All Cr (VI)-containing compounds were thought to be once made by man, with only Cr (III) naturally ubiquitous in air, water, soil, and biological materials. However, natural Cr (VI) has been recently found in ground and surface water at values exceeding the World Health Organization (WHO) limit for drinking water, that is of 50 μg of Cr (VI) per liter (Velma et al., 2009). Cr is widely used in numerous industrial processes and, as a result, it is a contaminant of many environmental systems (Cohen et al., 1993). Commercially, Cr compounds are used in industrial welding, chrome plating, dyes, pigments, leather tanning, and wood preservation. Cr is also used as an anticorrosive compound in cooking systems and boilers (Norseth, 1981; Wang et al., 2006).

It is estimated that more than 300,000 workers are annually exposed to Cr and Cr-containing compounds in their workplace. In humans and animals, [Cr (III)] is an essential nutrient that plays a role in the metabolism of glucose, fats, and proteins, by potentiating the action of insulin (WHO, 1996). However, occupational exposure has been a major concern due to the high risk of Cr-induced diseases in industrial workers professionally exposed to Cr (VI). In addition, general human population and some wild animals may also be at risk. It is estimated that 33 tons of total Cr are released annually into the environment (ATSDR, 2012b). The Occupational Safety and Health Administration of the United States (OSHA) recently established a "safe" level of 5 $\mu g/m^3$, for a time-weighted average of 8 h, although this revised level may still represent a carcinogenic risk (Zhang et al., 2011). For general human population, atmospheric levels range from 1 to 100 ng/cm^3 (Singh et al., 1999), but this range may be exceeded in areas closer to Cr manufacture. Non-occupational exposure occurs through Cr ingestion by contaminated food and water, while occupational exposure occurs by inhalation (Langård and Vigander, 1983). Range of Cr concentrations are between 1 and 3000 mg/kg in soil, 5 and 800 µg/L in seawater, and 26 µg/L and 5.2 mg/L in rivers and lakes (Jacobs and Testa, 2005). Cr content in food may vary a lot depending on the preparation process. In general, most fresh foods usually contain Cr levels ranging from <10 to 1300 µg/kg, and current workers in the Cr-related industries may be exposed to Cr concentrations two orders of magnitude higher than the general population (ATSDR, 2012b). Although the main route of Cr human exposure is inhalation, and the lung is the main target organ, significant Cr exposure through the skin has also been reported (Costa, 1997; Shelnutt et al., 2007). For example, the widespread incidence of dermatitis noted among construction workers is attributed to Cr exposure present in cement (Shelnutt et al., 2007). It is known that occupational and environmental exposure to Cr (VI) containing compounds causes multi-organ toxicity, such as kidney damage, allergy, asthma, and cancer of the respiratory tract in humans (Paustenbach et al., 1991).

An increase in stomach tumors has been observed in humans and animals exposed to Cr (VI) contained in drinking water. Accidental or intentional ingestion of extremely high doses of Cr (VI) compounds by humans has led to respiratory, cardiovascular, gastrointestinal, hematological, hepatic, renal, and severe neurological effects as part of the sequelae that led to death, or in patients who survived due to medical treatment (Costa, 1997). Although the evidence of Cr carcinogenicity in humans and terrestrial mamma seems strong, the mechanism by which it causes cancer is not fully understood (Chen et al., 2009).

2.5.4 Lead (Pb)

Pb is a metal of natural origin present in small quantities on the Earth's crust. Although Pb occurs naturally in the environment, anthropogenic activities such as burning fossil fuels, mining and manufacturing contribute to its release in high concentrations. Pb has many different industrial, agricultural, and domestic applications. It is currently used in the production of Pb-acid batteries, ammunition, metal products (welding and tubes), and X-ray shielding devices. It is estimated that 1.52 million metric tons of Pb were used for various industrial applications in the U.S in 2004. Of that amount, production of Pb batteries accounted for 83%, and the remaining percentage was covered by the fabrication of a range of products, such as ammunition (3.5%), oxides for paint, glass, pigments and chemicals (2.6%), and Pb in sheet (1.7%) (Gabby, 2003, 2006). In recent years, the industrial use of

Pb has been significantly reduced from paints and ceramic products and from caulking and pipe welding (CDC, 1985). Despite this progress, it is reported that out of 16.4 million households in the U.S with more than one child under the age of 6 per household, 25% of them still had significant amounts of deteriorated paint, dust, or adjacent bare soil contaminated with Pb (Jacobs et al., 2002), and this Pb in dust and soil often recontaminates cleaned houses (Farfel and Chisolm, 1991), also contributing to elevated blood Pb concentrations in children who play on this bare, contaminated soil (CDC, 2002). Today, the greatest source of Pb poisoning in children comes from dust and chips from the deterioration of Pb paint on interior surfaces (Lanphear et al., 1998). Children living in homes with deteriorated Pb paint may reach blood Pb concentrations of 20 μg/dL or greater (Charney et al., 1980). Exposure to Pb occurs mainly by inhalation of contaminated dust particles or aerosols and ingestion of food, water, and Pb-contaminated paints (CDC, 1992; ATSDR, 2007). Adults absorb 35%–50% of Pb through drinking water and the absorption rate for children can be greater than 50%. This Pb absorption is influenced by factors such as age and physiological state. In the human body, the highest percentage of Pb is carried to the kidney, followed by the liver and other soft tissues, such as the heart and brain. However, Pb in the skeleton represents the largest fraction of the body (Flora et al., 2006). The nervous system is the most vulnerable target of Pb poisoning. Headache, poor attention, irritability, loss of memory, and dullness are the first symptoms of Pb exposure effects in the central nervous system (CDC, 2002; ATSDR, 2007).

Since the late 1970s, Pb exposure has decreased significantly as a result of multiple efforts, including the elimination of Pb in gasoline and the reduction of its levels in the residential sector, like in paints, food, beverage cans, and piping systems (CDC, 1992; ATSDR, 2007). Several federal programs implemented by state and local health governments have not only focused on banning Pb in gasoline, paint, and soldered cans, but have also supported programs for detecting Pb poisoning in children and reducing it in homes (CDC, 1985). Despite the progress in these programs, Pb human exposure remains a serious health problem (Prikle et al., 1994; 1998). Pb is the largest of the systemic toxins that affect various organs of the body, including the kidneys, liver, central nervous system, hematopoietic system, endocrine system, and reproductive system (ATSDR, 2007).

There are many published studies that have documented the adverse effects of Pb in children and adult population. In children, these studies have shown an association between the level of blood poisoning and decreased intelligence, lower IQ, delayed or altered neurobehavioral development, decreased auditory acuity, speech and language disabilities, growth retardation, lack of attention, social incapacity and diligent behavior (Litvak et al., 1998; Kaul et al., 1999; EPA, 2002). In the adult population, reproductive effects such as decreased sperm count in men and spontaneous abortions in women have been associated with high Pb exposure (Amodio-Cocchieri et al., 1996; Litvak et al., 1998). Acute Pb exposure induces brain damage, kidney damage, and gastrointestinal diseases, while chronic exposure can cause adverse effects on blood, central nervous system, blood pressure, kidneys, and vitamin D metabolism (CDC, 1992; Amodio-Cocchieri et al., 1996; Apostoli et al., 1998; Litvak, 1998; Kaul et al., 1999; Hertz-Picciotto, 2000; EPA, 2002; ATSDR, 2007).

2.5.5 Mercury (Hg)

Hg is a heavy metal that belongs to the series of transition elements of the periodic table. It is unique because it exists or is found in nature in three forms (elemental, inorganic, and organic), each one with its own toxicity profile (Clarkson et al., 2003). At room temperature, elemental Hg exists as a liquid that has a high vapor pressure and is released into the environment as Hg vapor. Hg also exists as a cation with oxidation states of $+1$ (mercury) or $+2$ (mercuric) (Guzzi and LaPorta, 2008). Methylmercury (MeHg) is the most frequent compound of the organic form found in the environment and is formed as a result of the methylation of inorganic (mercuric) Hg forms by microorganisms found in soil and water (Dopp et al., 2004). Hg is a toxic and widespread environmental pollutant that induces serious alterations in body tissues and causes a wide range of adverse health effects (Bhan and Sarkar, 2005).

Humans and animals are exposed to various Hg chemical forms in the environment. These include elemental Hg vapor (Hg^0), inorganic Hg (Hg^{+1}), mercuric (Hg^{+2}), and organic Hg compounds (Zahir et al., 2005). Because Hg is everywhere in the environment, humans, plants, and animals cannot avoid exposure to some Hg form (Holmes et al., 2009). Hg is used in the electrical industry (switches, thermostats, and batteries), dentistry (dental amalgams), and numerous industrial processes, including the production of caustic soda, in nuclear reactors, as antifungal agents for wood processing, as a solvent for reagents and precious metals, and as a preservative of pharmaceutical products (Tchounwou et al., 2003a). Industrial Hg demand peaked in 1964 and began to decline sharply between 1980 and 1994 as a result of federal bans on Hg additives in paints, pesticides, and reduced battery use(EPA, 1997).

Humans are exposed to all forms of Hg through accidents, environmental contamination, food contamination, dental care, preventive medical practices, industrial and agricultural operations, and occupational operations (Bhan and Sarkar, 2005). Dental amalgams and fish consumption are the main sources of exposure to Hg chronic low levels. Hg enters water as a natural process of the Earth's crust degassing and also through industrial pollution (Dopp et al., 2004). Algae and bacteria methylate the Hg that enters the waterways. MeHg then makes its way through the food chain into fish, shellfish, and finally into humans (Sanfeliu et al., 2003). The two most highly absorbed species are elemental Hg (Hg^0) and MeHg. Dental amalgams contain more than 50% of elemental Hg (Zahir et al., 2005). Elemental vapor is highly lipophilic and is efficiently absorbed through the lungs and tissues lining the mouth. After Hg^0 enters the blood, it rapidly passes through the cell membranes, which include both the blood–brain barrier and the placental barrier. Once inside the cell, the Hg^0 is oxidized and turns into the highly reactive Hg^{2+}. The MeHg derived from eating fish is easily absorbed in the gastrointestinal tract and, due to its solubility in lipids, can easily cross the placental and blood–brain barriers. Once the Hg is absorbed, it has a very low excretion rate. A significant proportion of what is absorbed accumulates in the kidneys, neurological tissue, and liver. All forms of Hg are toxic and their effects include gastrointestinal toxicity, neurotoxicity, and nephrotoxicity (Guzzi and LaPorta, 2008).

2.6 Conclusions

Based on all information to date, metallic compounds and high metal content residues represent a serious environmental and health threaten when present in ecosystems, as metals get accumulated and are hard to remove. Thus, the development of new approaches is imminent to prevent their usage and production and to treat and remove this kind of components before they impact water bodies, soil, and air. Also, it is important to improve accurate detection systems that may determine their presence and speciation under different conditions, so their impact may be mitigated.

References

Agency for Toxic Substances and Disease Registry – ATSDR 2000. Toxicological profile for Arsenic TP-92/09. Georgia: Center for Disease Control, Atlanta, U.S. https://www.atsdr.cdc.gov/toxprofiles/tp.asp?id=22&tid=3

Agency for Toxic Substances and Disease Registry Agency – ATSDR 2007. Toxicological profile for lead. Center for Disease Control, Department of Health and Human Services, Public Health Service. Atlanta, GA., U.S. https://www.atsdr.cdc.gov/toxprofiles/tp.asp?id=96&tid=22

Agency for Toxic Substances and Disease Registry – ATSDR 2012a. Toxicological profile for cadmium. Atlanta, GA, U.S. https://www.atsdr.cdc.gov/toxprofiles/tp.asp?id=48 &tid=15

Agency for Toxic Substances and Disease Registry Agency – ATSDR 2012b. Toxicological profile for chromium. Center for Disease Control, Department of Health and Human Services, Public Health Service. Atlanta, GA., U.S. https://www.atsdr.cdc.gov/toxprofiles/tp7.pdf

Akesson, A., Bjellerup, P., Lundh, T., Lidfeldt, J., Nerbrand, C., Samsioe, G., et al., 2006. Cadmium-induced effects on bone in a population-based study of women. Environ. Health Perspect., 114, 830–834. https://doi.org/10.1289/ehp.8763

Amodio-Cocchieri, R., Arnese, A., Prospero, E., Roncioni, A., Barulfo, L., Ulluci, R. et al., 1996. Lead in human blood form children living in Campania, Italy. J. Toxicol. Environ. Health, 47, 311–320. https://doi.org/10.1080/009841096161663

Ancheyta, J., Speight. J.G., 2007. Hydroprocessing of heavy oils and residua, Chemical industries, CRC Press, ISBN: 9780849374197 – CAT# 7419.

Apostoli, P., Kiss, P., Stefano, P., Bonde, J.P., Vanhoorne M., 1998. Male reproduction toxicity of lead in animals and humans. Occupa. Environ. Med., 55, 364–374. https://www.ncbi.nlm.nih.gov/pmc/articles/PMC1757597/pdf/v055p00364.pdf

Baselt, R.C., 2017. Disposition of toxic drugs and chemicals in man. 11th ed. Year Book Medical Publishers, Biomedical Pubns. Chicago, IL, U.S. pp. 105–107. ISBN-10: 9780692774991.

Becker, K., Kaus, S., Krause, C., Lepom, P., Schulz, C., Seiwert, M., et al., 2002. German Environmental Survey 1998 (GerES III): Environmental pollutants in blood of the German population. Int. J. Hyg. Environ. Health, 205, 297–308. https://doi.org/10.1078/1438-4639-00155

Berkhout, F., Hertin, J., 2004. De-materialising and re-materialising: digital technologies and the environment. Futures, 36, 903–920. https://doi.org/10.1016/j.futures.2004.01.003

Bertram, M., Graedel, T.E., Rechberger, H., Spatari, S., 2002. The contemporary European copper cycle: Waste management subsystem. Ecol. Econ., 42, 43–57. https://doi.org/10.1016/S0921-8009(02)00100-3

Betts, K., 2008. Producing usable materials from e-waste. Environ. Sci. Technol., 42, 6782–6783. https://doi.org/10.1021/es801954d

Bhan, A., Sarkar, B.A., 2005. Mercury in the environment: Effects on health and reproduction. Rev. Environ. Health. 20, 39–56. https://doi.org/10.1515/REVEH.2005.20.1.39

Bradl, H., 2002. Sources and origins of heavy metals. In: Heavy metals in the environment: Origin, interaction and remediation. Vol. 6. Academic Press, London, U.K. ISBN: 9780080455006.

Centeno, J.A., Tchounwou, P.B., Patlolla, A.K., Mullick, F.G., Murakat, L., Meza, E., et al., 2006. Environmental pathology and health effects of arsenic poisoning: A critical review. In: Managing arsenic in the environment: From soil to human health. Naidu, R., Smith, E., Smith, J., Bhattacharya, P. (Eds.). CSIRO Publishing Corp., Adelaide, Australia.

Centers for Disease Control and Prevention – CDC 1985. Preventing lead poisoning in young children: A statement by the Centers for Disease Control. Morbidity and Mortality Weekly Report, 34, 66–8, 73. https://www.cdc.gov/mmwr/preview/mmwrhtml/00000659.htm

Centers for Disease Control and Prevention – CDC 1992. Case Studies in Environmental Medicine – Lead Toxicity. Atlanta, GA, U.S. https://wonder.cdc.gov/wonder/prevguid/p0000017/p0000017.asp

Centers for Disease Control and Prevention – CDC 2002. Managing elevated blood Lead levels among young children: Recommendations from the Advisory Committee on Childhood Lead Poisoning Prevention. Atlanta, GA, U.S. https://stacks.cdc.gov/view/cdc/26980

Charney, E., Sayre, J., Coulter M., 1980. Increased lead absorption in inner city children: where does the lead come from? Pediatrics, 6, 226–231. https://pediatrics.aappublications.org/content/65/2/232

Chen, T.L., Wise, S.S., Kraus, S., Shaffiey, F., Levine, K., Thompson, D.W. et al., 2009. Particulate hexavalent chromium is cytotoxic and genotoxic to the North Atlantic right whale (*Eubalaena glacialis*) lung and skin fibroblasts. Environmental and Molecular Mutagenesis, 50, 387–393. https://doi.org/10.1002/em.20471

Chikusa, C.M., 1994. Pollution caused by mine dumps and its control. Rhodes University. Grahamstown, South Africa. http://hdl.handle.net/20.500.11892/ 22 574

Chiranjeevi, T., Pragya, R., Gupta, S., Gokak, D.T., Bhargava, S., 2016. Minimization of waste spent catalyst in refineries. Procedia Environ. Sci., 35, 610–617. https://doi.org/10.1016/j.proenv.2016.07.047

Chon, H.T., Hwang, J.H., 2000. Geochemical characteristics of the acid mine drainage in the water system in the vicinity of the Dogye Coal Mine in Korea. Environ. Geochem. Health, 22, 155–172. https://doi.org/10.1023/A:1006735226263

Chrzan, A., 2016. Monitoring bioconcentration of potentially toxic trace elements in soils trophic chains. Environ. Earth Sci., 75, 786. https://doi.org/10.1007/s12665-016-5595-4

Clarkson, T.W., Magos, L., Myers, G.J., 2003. The toxicology of mercury-current exposures and clinical manifestations. The New England Journal of Medicine, 349, 1731–1737. https://doi.org/10.1056/NEJMra022471

Cobbing, M., 2008. Toxic Tech: Not in our backyard. Uncovering the hidden flows of e-waste. Report from Greenpeace International. Amsterdam, The Netherlands. http://www.greenpeace.org/raw/content/belgium/fr/press/reports/toxic-tech.pdf

Cohen, M.D., Kargacin, B., Klein, C.B., Costa, M., 1993. Mechanisms of chromium carcinogenicity and toxicity. Crit. Rev. Toxicol., 23, 255–281. https://doi.org/10.3109/10408449309105012

Costa, M., 1997. Toxicity and carcinogenicity of Cr (VI) in animal models and humans. Crit. Rev. Toxicol., 27, 431–442. https://doi.org/10.3109/10408449709078442

Cox, D.P., Wright, N.A., Coakley, G.J., 1981. The nature and use of copper reserve and resource data. U.S. Department of The Interior, United States Government Printing Office, Washington, U.S. https://pubs.usgs.gov/pp/0907f/report.pdf

Davison, A.G., Fayers, P.M., Taylor, A.J., Venables, K.M., Darbyshire, J., Pickering, C.A., et al., 1988. Cadmium fume inhalation and emphysema. The Lancet 331, 663–667. https://doi.org/10.1016/S0140-6736(88)91474-2

Dopp, E., Hartmann, L.M., Florea, A.M., Rettenmier, A.W., Hirner, A.V., 2004. Environmental distribution, analysis, and toxicity of organometal (loid) compounds. Crit. Rev. Toxicol., 34, 301–333. https://doi.org/10.1080/10408440490270160

Dudka, S., Adriano, D.C., 1997. Environmental impacts of metal ore mining and processing: A review. J. Environ. Qual., 26, 590–602. http://citeseerx.ist.psu.edu/viewdoc/download?doi=10.1.1.462.7000&rep=rep1&type=pdf

Duffus, J.H., 2002. Heavy metals – a meaningless term? (IUPAC Technical Report) Pure and Applied Chemistry, 74, 793–807. https://doi.org/10.1351/pac200274050793

Dufresne, P., 2007. Hydroprocessing catalysts regeneration and recycling. Applied Catalysis A: General. 322, 67–75. https://doi.org/10.1016/j.apcata.2007.01.013

Elinder, C.G., Järup, L., 1996. Cadmium exposure and health risks: Recent findings. Ambio, 25, 370–373. http://www.jstor.org/stable/4314494

Environmental Protection Agency – EPA 1976. The ecological effects of Arsenic emitted from non-ferrous smelters. Office of Toxic Substances, Washington, U.S. https://nepis.epa.gov/Exe/ZyNET.exe/9100AN95.txt?ZyActionD=ZyDocument&Client=EPA&Index=1976%20Thru%201980&Docs=&Query=&Time=&EndTime=&SearchMethod=1&TocRestrict=n&Toc=&TocEntry=&QField=&QFieldYear=&QFieldMonth=&QFieldDay=&UseQField=&IntQFieldOp=0&ExtQFieldOp=0&XmlQuery=&File=D%3A%5CZYFILES%5CINDEX%20DATA%5C76THRU80%5CTXT%5C00000012%5C9100AN95.txt&User=ANONYMOUS&Password=anonymous&SortMethod=h%7C&MaximumDocuments=1&FuzzyDegree=0&ImageQuality=r75g8/r75g8/x150y150g16/i425&Display=hpfr&DefSeekPage=x&SearchBack=ZyActionL&Back=ZyActionS&BackDesc=Results%20page&MaximumPages=1&ZyEntry=1&slide

Environmental Protection Agency – EPA 1997. Mercury study report to Congress. https://www.epa.gov/mercury/mercury-study-report-congress

Environmental Protection Agency – EPA 2000. Cadmium Compounds. https://www.epa.gov/sites/production/files/2016-09/documents/cadmium-compounds.pdf

Environmental Protection Agency – EPA 2002. Lead Compounds. Technology Transfer Network – Air Toxics Website. http://www.epa.gov/cgi-bin/epaprintonly.cgi

Environmental Protection Agency – EPA 2003. Hazardous waste management system: Petroleum refining process wastes; Identification of characteristically hazardous self-heating solids; land disposal restrictions: Treatment standards for spent hydrorefining catalyst (K172) hazardous waste. Federal Register, 68(202), 59935–59940. https://www.federalregister.gov/documents/2003/10/20/03-26411/hazardous-waste-management-system-petroleum-refining-process-wasteside ntification-of

Environmental Protection Agency – EPA 2016. NAAQS Table. https://www.epa.gov/criteria-air-pollutants/naaqs-table#1

Environmental Protection Agency – EPA 2018a. Criteria for the definition of solid waste and solid and hazardous waste exclusions. https://www.epa.gov/hw/criteria-definition-solid-waste-and-solid-and-hazardous-waste-exclusions#tables w

Environmental Protection Agency – EPA 2018b. TENORM: Copper mining and production wastes. Radiation Protection. https://www.epa.gov/radiation/tenorm-copper-mining-and-production-wastes

Environmental Protection Agency – EPA 2019. Part 273 – Standards for universal waste management. Electronic Code of Federal Regulations. https://www.ecfr.gov/cgibin/textidx?SID=0501d91ec562faafa833c60c2404d806&c=true&node=pt40.27.273&rgn=div5

Farfel, M.R., Chisolm, J.J. Jr., 1991. An evaluation of experimental practices for abatement of residential lead-based paint: report on a pilot project. Environ. Res., 55, 199–212. https://doi.org/10.1016/S0013-9351(05)80176-8

Fergusson, J.E., 1990. The heavy elements: Chemistry, environmental impact and health effects. Pergamon Press, Oxford, UK. https://doi.org/10.1016/0269-7491(91)90124-F

Fernandes, C., Takeshi, L., Lopes, E.M., Omori, W.P., Souza J.A.M., Alves L.M.C., Lemos, E.G.M., 2018. Bacterial communities in mining soils and surrounding areas under regeneration process in a former ore mine. Environ. J. Microbiol., 49, 489–502. https://doi.org/10.1016/j.bjm.2017.12.006

Flora, S.J.S., Flora, G., Saxena, G., 2006. Environmental occurrence, health effects and management of lead poisoning. In: Lead: chemistry, analytical aspects, environmental impacts and health effects. Cascas, S.B., Sordo, J. (Eds.). Elsevier, Netherlands. pp. 158–228. https://doi.org/10.1016/B978-044452945-9/50004-X

Fowler, B.A., Chou, S., Jones, R.L., Chen, C.J., 2007. Arsenic. In: Handbook on the toxicology of metals, 3rd. edition. Nordberg, G.F., Fowler, B.A., et al. (Eds.), Elsevier Inc., pp. 367–406.

Furimsky, E., 2007. Catalysts for upgrading heavy petroleum feeds: Studies in surface science and catalysis, Vol. 169. Elsevier, Oxford, UK. ISBN: 9780080549316.

Gabby, P.N., 2003. Lead, Environmental Defense, Alternatives to Lead-Acid Starter Batteries, Pollution Prevention Fact Sheet. http://www.cleancarcampaign.org/FactSheet_BatteryAlts.pdf

Gabby, P.N., 2006. Lead. In: Mineral Commodity Summaries, Geological Survey. Reston, VA, U.S. http://minerals.usgs.gov/minerals/pubs/commodity/lead/lead_mcs05.pdf

Gallagher, C.M., Kovach, J.S., Meliker, J.R., 2008. Urinary cadmium and osteoporosis in U.S. women $\geq$50 years of age: NHANES 1988–1994 and 1999–2004. Environ. Health Perspect., 116, 1338–1343. https://doi.org/10.1289/ehp.11452

Goyer, R.A., 2001. Toxic effects of metals. In: Cassarett and Doull's Toxicology: The basic science of poisons. Klaassen, C.D. (Ed.) McGraw-Hill, New York, U.S. pp. 811–867.

Guzzi, G., LaPorta, C.A., 2008. Molecular mechanisms triggered by mercury. Toxicology, 244, 1–12. https://doi.org/10.1016/j.tox.2007.11.002

Hailemariam, M., Mushir, A., Zbelo, T., Niguse, A., 2017. Impacts of artisanal gold mining systems on soil and woody vegetation in the semi-arid environment of northern Ethiopia. Singap. J. Trop. Geogr., 38, 386–401. https://doi.org/10.1111/sjtg.12203

He, Z.L., Yang, X.E., Stoffella, P.J., 2005. Trace elements in agroecosystems and impacts on the environment. J. Trace Elem. Med. Biol., 19, 125–140. https://doi. org/10.1016/j.jtemb.2005.02.010

Herawati, N., Suzuki, S., Hayashi, K., Rivai, I.F., Koyoma, H., 2000. Cadmium, copper and zinc levels in rice and soil of Japan, Indonesia and China by soil type. Bull. Environ. Contam. Toxicol., 64, 33–39. https://doi.org/10.1007/s001289910006

Hertz-Picciotto, I., 2000. The evidence that lead increases the risk for spontaneous abortion. Am. J. Ind. Med., 38, 300–309.

Hoffmann, J.E., 1992. Recovering precious metals from electronic scrap. The Journal of The Minerals, Metals & Materials Society, 44, 43–48. https://doi.org/10.1007/BF03222275

Holmes, P., James, K.A., Levy, L.S., 2009. Is low-level mercury exposure of concern to human health? Sci. Total Environ., 408, 171–182. https://doi.org/10.1016/j.scitotenv.2009.09.043

International Agency for Research on Cancer – IARC 1990. Chromium, nickel and welding. Monographs on the evaluation of carcinogenic risks to humans. Monographs – Vol. 49. pp. 677. https://monographs.iarc.fr/wp-content/uploads/2018/06/mono49.pdf

International Agency for Research on Cancer – IARC 2018. Cadmium and cadmium compounds. Monographs – 100C. pp. 121–145. https://monographs.iarc.fr/wp-content/uploads/2018/06/mono100C-8.pdf

Jacobs, D.E., Clickner, R.P., Zhou, J.Y., Viet, S.M., Marker, D.A., Rogers, J.W., et al., 2002. The prevalence of lead-based paint hazards in U.S. housing. Environ. Health Perspect., 110, A599–A606. https://doi.org/10.1289/ehp.021100599

Jacobs, J.A., Testa, S.M., 2005. Overview of chromium (VI) in the environment: Background and history. In: Chromium (VI) Handbook. Guertin, J., Jacobs, J.A., Avakian, C.P. (Eds.). CRC Press, Boca Raton, FL, U.S. pp. 1–22. https://doi.org/10.1017/S1047951104006225

Järup, L., Berglund, M., Elinder, C.G., Nordberg, G., Vahter, M., 1998. Health effects of cadmium exposure – a review of the literature and a risk estimate. Scand. J. Work Environ. Health, 24, 1–51. http://www.sjweh.fi/show_abstract.php?abstract_id=281

Joint Group of Experts on the Scientific Aspects of Marine Environmental Protection – GESAMP 1987. Report of the 17th session. Reports and Studies No. 31, Geneva, Switzerland. http://www.gesamp.org/data/gesamp/files/media/Publications/Reports_and_studies_31/gallery_1264/object_1272_large.pdf

Karrari, P., Mehrpour, O., Abdollahi, M., 2012. A systematic review on status of lead pollution and toxicity in Iran; Guidance for preventive measures. DARU Journal of Pharmaceutical Sciences. 10, 1–17. https://doi.org/10.1186/1560-8115-20-2

Kaul, B., Sandhu, R.S., Depratt, C., Reyes, F., 1999. Follow-up screening of lead-poisoned children near an auto battery recycling plant, Haina, Dominican Republic. Environ. Health Perspect., 107, 917–920. https://doi.org/10.1289/ehp.99107917

LaDou, J., Lovegrove, S., 2008. Export of electronics equipment waste. Int. J. Occup. Environ. Health, 14, 1–10. https://doi.org/10.1179/oeh.2008.14.1.1

Langård, S., Vigander T., 1983. Occurrence of lung cancer in workers producing chromium pigments. Br. J. Ind. Med., 40, 71–74. https://www.ncbi.nlm.nih.gov/pmc/articles/PMC1009121/pdf/brjindmed00053-0075.pdf

Lanphear, B.P., Matte, T.D., Rogers, J., Clickner, R.P., Dietz, B., Bornschein, R.L., et al., 1998. The contribution of lead-contaminated house dust and residential soil to children's blood lead levels. A pooled analysis of 12 epidemiologic studies. Environ. Res., 79, 51–68. https://doi.org/10.1006/enrs.1998.3859

Litvak, P., Slavkovich, V., Liu, X., Popovac, D., Preteni, E., Capuni-Paracka, S. et al., 1998. Hyperproduction of erythropoietin in nonanemic leadexposed children. Environ. Health Perspect., 106, 361–364. https://doi.org/10.1289/ehp.98106361

Lu, C., Zhang, L., Zhong, Y., Ren, W., Tobias, M., Mu, Z., et al., 2015. An overview of e-waste management in China. Journal of Material Cycles and Waste Management, 17, 1–12. http://dx.doi.org/10.1007/s10163-014-0256-8

Lua, Z., Caib, M., 2012. Disposal methods on solid wastes from mines in transition from open-pit to underground mining. The 7th International Conference on Waste Management and Technology. Procedia Environ. Sci., 16, 715–721. https://doi.org/10.1016/j.proenv.2012.10.098

Luo, C., Liu, C., Wang, Y., Liu, X., Li, F., Zhang, G., et al., 2011. Heavy metal contamination in soils and vegetables near an e-waste processing site, south China. J. Hazard. Mater., 186, 481–490. https://doi.org/10.1016/j.jhazmat.2010.11.024

Mannino, D.M., Holguin, F., Greves, H.M., Savage-Brown, A., Stock, A.L., Jones, R.L., 2004. Urinary cadmium levels predict lower lung function in current and former smokers: data from the Third National Health and Nutrition Examination Survey. Thorax, 59, 194–198. https://doi.org/0.1136/thorax.2003.012054

Mascagni, P., Consonni, D., Bregante, G., Chiappino, G., Toffoletto F., 2003. Olfactory function in workers exposed to moderate airborne cadmium levels. NeuroToxicology, 24, 717–724. https://doi.org/10.1016/S0161-813X(03)00024-X

Morton, W.E., Dunnette, D.A., 1994. Health effects of environmental arsenic. In: Arsenic in the environment Part II: Human health and ecosystem effects. Nriagu, J.O. (Ed.). John Wiley & Sons, Inc. New York, U.S., pp. 17–34.

National Research Council – NRC 2001. Arsenic in Drinking Water. Division on Earth and Life Studies; Board on Environmental Studies and Toxicology; Subcommittee to Update the 1999 Arsenic in Drinking Water Report; Committee on Toxicology. Washington, D.C., U.S. https://www.nap.edu/catalog/10194/arsenic-in-drinking-water-2001-update

National Research Council Canada – NRCC 1978. Effects of Arsenic in the Environment.

Ni, H.G., Zeng, E.Y., 2009. Law enforcement and global collaboration are the keys to containing e-waste tsunami in China. Environ. Sci. Technol., 43, 3991–3994. https://doi.org/10.1021/es802725m

Nordberg, G.F., Fowler, B.A., Nordberg, M., 2015. Handbook on the toxicology of metals. Elsevier Inc. https://doi.org/10.1016/C2011-0-07884-5

Norseth, T., 1981. The carcinogenicity of chromium. Environ. Health Perspect., 40, 121–130. https://www.ncbi.nlm.nih.gov/pmc/articles/PMC1568823/pdf/envhper 00467-0124.pdf

Paschal, D.C., Burt, V., Caudill, S.P., Gunter, E.W., Pirkle, J.L., Sampson, E.J., et al., 2000. Exposure of the U.S. population aged 6 years and older to cadmium: 1988–1994. Arch. Environ. Contam. Toxicol., 38, 377–383. https://doi.org/10.1007/s002449910050

Paustenbach, D.J., Rinehart, W.E., Sheehan, P.J., 1991. The health hazards posed by chromium-contaminated soils in residential and industrial areas: Conclusions of an expert panel. Regul. Toxicol. Pharmacol., 13, 195–222. https://doi.org/10.1016/0273-2300(91)90022-N

Pirkle, J.L., Brady, D.J., Gunter, E.W., Kramer, R.A., Paschal, D.C., Flegal, K.M. et al., 1994. The decline in blood lead levels in the United States: The National Health and Nutrition Examination Surveys (NHANES). Journal of the American Medical Association, 272, 284–291. https://doi.org/10.1001/jama.1994.03520040046039

Pirkle, J.L., Kaufmann, R.B., Brody, D.J., Hickman, T., Gunter, E.W., Paschal, D.C., 1998. Exposure of the U.S. population to lead: 1991–1994. Environ. Health Perspect., 106, 745–750. https://doi.org/10.1289/ehp.98106745

Rashid Khan M., Sayed E., 2011. Sulfur removal from heavy and light petroleum hydrocarbon by selective oxidation. Advances in Clean Hydrocarbon Fuel Processing. 2011, 243–261. https://doi.org/10.1533/9780857093783.3.243

Ratnaike, R.N., 2003. Acute and chronic arsenic toxicity. Postgrad. Med. J., 79, 391–396. http://dx.doi.org/10.1136/pmj.79.933.391

Robinson, B.H., 2009. E-waste: an assessment of global production and environmental impacts. Sci. Total Environ., 408, 183–191. https://doi.org/10.1016/j.scitotenv.2009.09.044

Sanfeliu, C., Sebastia, J., Cristofol, R., Rodriquez-Farre E., 2003. Neurotoxicity of organomercurial compounds. Neurotox. Res., 5, 283–305. https://doi.org/10.1007/BF03033386

Satarug, S., Baker, J.R., Urbenjapol, S., Haswell-Elkins, M., Reilly, P.E., Williams, D.J., et al., 2003. A global perspective on cadmium pollution and toxicity in non-occupationally exposed population. Toxicol. Lett., 137, 65–83. https://doi.org/10.1016/S0378-4274(02)00381-8

Schmidt, C.W., 2002. E-junk explosion. Environ Health Perspect., 110, A188–194. https://doi.org/10.1289/ehp.110-a188

Schutte, R., Nawrot, T.S., Richart, T., Thijs, L., Vanderschueren, D., Kuznetsova, T., et al., 2008. Bone resorption and environmental exposure to cadmium in women: a population study. Environ. Health Perspect., 116, 777–783. https://doi.org/10.1289/ehp.11167

Sclzaefer, M.O., Gutzmer, J., Beukes, N., 2004. Mineral chemistry of sphalerite and galena from Pb-Zn mineralization hosted by the transvaal supergroup in Griqualand west, South Africa. Transactions – Geological Society of South Africa, 10, 341–354.

Shallari, S., Schwartz, C., Hasko, A., Morel, J.L., 1998. Heavy metals in soils and plants of serpentine and industrial sites of Albania. Sci. Total Environ., 19, 133–142. https://doi.org/10.1016/S0048-9697(98)80104-6

Shelnutt, S.R., Goad, P., Belsito, D.V., 2007. Dermatological toxicity of hexavalent chromium. Crit. Rev. Toxicol., 37, 375–387. https://doi.org/10.1080/10408440701266582

Singh, J., Pritchard, D.E., Carlisle, D.L., Mclean, J.A., Montaser, A., Orenstein, J.M., et al., 1999. Internalization of carcinogenic lead chromate particles by cultured normal human lung epitelial cells: Formation of intracellular lead-inclusion bodies and induction of apoptosis. Toxicol. Appl. Pharmacol., 161, 240–248. https://doi.org/10.1006/taap.1999.8816

Singh, R., Gautam, N., Mishra, A., Gupta, R., 2011. Heavy metals and living systems: An overview. Indian J. Pharmacol., 43, 246–253. https://doi.org/10.4103/0253-7613.81505

Terraglio, F.P., Manganelli, R.M., 1967. The absorption of atmospheric sulfur dioxide by water solutions. J. Air Pollut. Control Assoc. 17, 403–406. https://doi.org/10.1080/00022470.1967.10468999

Tchounwou, P.B., Ayensu, W.K., Ninashvilli, N., Sutton, D., 2003a. Environmental exposures to mercury and its toxicopathologic implications for public health. Environ. Toxicol., 18, 149–175. https://doi.org/10.1002/tox.10116

Tchounwou, P.B., Patlolla, A.K., Centeno, J.A., 2003b. Carcinogenic and systemic health effects associated with arsenic exposure – a critical review. Toxicol. Pathol., 31, 575–588. https://doi.org/10.1080/01926230390242007

Tchounwou, P.B., Wilson, B., Ishaque, A., 1999. Important considerations in the development of public health advisories for arsenic and arsenic-containing compounds in drinking water. Rev. Environ. Health, 14, 211–229.

Tchounwou, P.B., Yedjou, C.G., Patlolla, A.K., Sutton, D.J., 2012. Heavy metals toxicity and the environment. EXS, 101, 133–164. https://doi.org/10.1007/978-3-7643-8340-4_6

United Nations Environment Programme – UNEP 2009. Basel Convention on the control of transboundary movements of hazardous wastes and their disposal. https://www.basel.int/portals/4/basel%20convention/docs/text/baselconventiontext-e.pdf

Velma, V., Vutukuru, S.S., Tchounwou, P.B., 2009. Ecotoxicology of hexavalent chromium in freshwater fish: A critical review. Rev. Environ. Health, 24, 129–145. https://www.ncbi.nlm.nih.gov/pmc/articles/PMC2860883/pdf/nihms-190522.pdf

Waalkes, M.P., 2000. Cadmium carcinogenesis in review. J. Inorg. Biochem., 79, 241–244. https://doi.org/10.1016/S0162-0134(00)00009-X

Waalkes, M.P., Rehm S., 1992. Carcinogenicity of oral cadmium in the male Wistar (WF/NCr) rat: effect of chronic dietary zinc deficiency. Fundaments of Applied Toxicology, 19, 512–520.

Wang, X.F., Xing, M.L., Shen, Y., Zhu, X., Xu, L.H. (2006). Oral administration of Cr (VI) induced oxidative stress, DNA damage and apoptotic cell death in mice. Toxicology, 228, 16–23. https://doi.org/10.1016/j.tox.2006.08.005

Wilmoth, R.C., Hubbard, S.J., Bruckle, J.O., Martin, J.F., 1991. Production and processing of metals: their disposal and future risks. In: Metals and their compounds in the environment. Merian, E. (Ed.). VCH: Weinheim, pp. 19–65.

Wittman, R., Hu, H., 2002. Cadmium exposure and nephropathy in a 28-year-old female metals worker. Environ. Health Perspect., 110, 1261–1266. https://doi.org/10.1289/ehp.021101261

World Health Organization – WHO 1996. Trace Elements in Human Nutrition and Health. Geneva, Switzerland. https://www.who.int/nutrition/publications/micro nutrients/9241561734/en/

World Health Organization – WHO 2017. Recycling used lead-acid batteries: health considerations. WHO, Switzerland. https://apps.who.int/iris/bitstream/handle/10665/259447/9789241512855-eng.pdf;jsessionid=A06F3EA8158BA63465FD E55A639C0586?sequence=1

World Health Organization – WHO 2018. Arsenic. https://www.who.int/news-room/fact-sheets/detail/arsenic

Yang, Z.Z., Zhao, X.R., Zhao, Q., Qin, Z.F., Qin, X.F., Xu, X.B., et al. 2008. Polybrominated diphenyl ethers in leaves and soil from typical electronic waste polluted area in South China. Bull. Environ. Contam. Toxicol., 80, 340–344. https://doi.org/10.1007/s00128-008-9385-x

Zahir, A., Rizwi, S.J., Haq, S.K., Khan, R.H., 2005. Low dose mercury toxicity and human health. Environ. Toxicol. Pharmacol., 20, 351–360. https://doi.org/10.1016/j.etap.2005.03.007

Zhang, J., Min H., 2009. Eco-toxicity and metal contamination of paddy soil in an e-wastes recycling area. J. Hazard. Mater., 165, 744–750. https://doi.org/10.1016/j.jhazmat.2008.10.056

Zhang, X.H., Zhang, X., Wang, X.C., Jin, L.F., Yang, Z.P., Jiang, C.X., et al., 2011. Chronic occupational exposure to hexavalent chromium causes DNA damage in electroplating workers. BMC Public Health, 11, 224. https://doi.org/10.1186/1471-2458-11-224

Zheng, J., Chen, K.H., Yan, X., Chen, S.J., Hu, G.C., Peng, X.W., et al., 2013. Heavy metals in food, house dust, and water from an e-waste recycling area in South China and the potential risk to human health. Ecotoxicol. Environ. Saf., 96, 205–212. https://doi.org/10.1016/j.ecoenv.2013.06.017

3

Parameters Involved in Biotreatment of Solid Wastes Containing Metals

Marlenne Gómez Ramírez[1,*] and Sergio A. Tenorio Sánchez[2]

[1]Departamento de Biotecnología, Centro de Investigación en Ciencia Aplicada y Tecnología Avanzada del Instituto Politécnico Nacional, Cerro Blanco 141, Colinas del Cimatario, 76090 Santiago de Querétaro, Querétaro, México
[2]Departamento de Microbiología, Escuela Nacional de Ciencias Biológicas, Instituto Politécnico Nacional, Prolongación de Carpio y Plan de Ayala s/n, Col. Santo Tomás, 11340, Ciudad de México, México
E-mail: mgomezr@ipn.mx
*Corresponding Author

3.1 Introduction

There are several methods for the treatment of hazardous wastes and their technologies are based on two main approaches: either hydrometallurgy or pyrometallurgy. With the hydrometallurgical approach, metals are leached using acids or bases, while pyrometallurgy uses a heat treatment, such as roasting and smelting (Beolchini et al., 2010). Biotechnology finds application in treatment of air emissions, solid, and liquid wastes by various biological methods. Biological methods are considered viable environmental-friendly technologies and have been developed in the last years and associated with lower cost and energy requirements, in comparison to non-biological processes (Gómez-Ramírez et al., 2018a). In those methods, it is important to find suitable microorganisms to degrade organic substances under favorable conditions to complete the treatment. The advantages of the biotechnological treatment of hazardous wastes are biodegradation or detoxification of a wide variety of hazardous substances using natural microorganisms as well as the availability of a wide range of biotechnological methods for the total destruction of these wastes without the production of secondary hazardous derivatives. However, to intensify the biological treatment, it is a necessary requirement to add nutrients and acceptors of electrons as well as to control the optimal conditions. Thus, biotechnology provides a solution for the ecological degradation of harmful heavy metals and toxic chemicals (Sood and Chitre, 2017). The metabolism of each microorganism determine how it interact with the environment, with the contaminating agent and how they adapt and respond in presence of solid wastes. The biodegradation rate depends on different physical and chemical factors and microorganisms used.

Some factors involved are (i) microbe–metal interactions and microorganism used, (ii) growth medium composition and appropriate level of nutrients, (iii) chemical and metal composition of solid wastes, (iv) pulp density, (v) size of particle of solid waste, (vi) pH, (vii) temperature, (viii) inocula size, (ix) contact time, (x) oxygen, (xi) physiological state of the microorganisms (culture age), (xii) microbial tolerance, (xiii) oxidation state of metals, and (xiv) presence of other toxic compounds (Das et al., 2008; Dzionek et al., 2016; Girma, 2015a; Gómez-Ramírez et al., 2014a, 2018a; Goyal et al., 2003; Karigar and Rao, 2011; Ojuederie and Babalola, 2017; Pradhan et al., 2009; Zhuang et al., 2015). In recent years, the mining industry has made significant efforts to develop eco-friendly and low-cost biohydrometallurgical operations. However, certain bottlenecks still exist which hinder its wider commercial applicability. The kinetic of the process is currently much too slow for it to be economical. Long periods of operation are required, compared to traditional methods of leaching to obtain reasonable yields (Pathak et al., 2017). The main purpose of this chapter is to present and discuss the parameters involved in biotreatment of solid waste containing metals.

3.2 Microbe–metal Interactions and Microorganisms Used

Microbial growth, differentiation, and metabolism are affected directly or indirectly by metals. The microorganisms are affected in different ways which depend on the type of metal or metallics compound, the environment, and organism, while structural components and metabolic activity also influence metal speciation and therefore solubility, mobility, bioavailability, and toxicity (Brandl and Faramarzi, 2006; Gadd, 2010). Many metals are essential for life, e.g., Ca, Zn, Fe, Mn, Na, Co, Mg, K, and Cu, but they can become toxic when they exceed certain threshold concentrations. There is no knowledge of essential metabolic functions of other metals such as Al, Cs, Hg, Pb, and Cd, but all can be bioaccumulated (Gadd, 2010). In the presence of mineral and metallic compounds in solid wastes, several metabolic reactions (metal trans-formations) can occur when microorganisms are in contact with solid metals or metals in solution. Mechanisms resulting in *mobilization of solid metals* such as: (a) redoxolysis, (b) acidolysis, (c) complexolysis, and (d) alkylation or *immobilization of solubilized metals* as: (a) biosorption, (b) bioaccumula-tion, (c) redox reaction, and (d) complex formation (Brandl and Faramarzi, 2006). Autotrophic and heterotrophic bacteria, and heterotrophic fungi are commonly used in biotreatment of solid wastes contaminated with metals (Abdullah et al., 2018; Adebayo and Obiekezie, 2018).

The use of autotrophic bacteria is advantageous because an organic car-bon source is not needed for their growth. On the other hand, heterotrophic bacteria and fungi can be used at higher pHs (i.e., alkaline and acid-consuming materials) (Adebayo and Obiekezie, 2018; Asghari et al., 2013). Bacteria from the genus *Leptospirillum*, *Acidithiobacillus*, and *Sulfolobus* are the most important autotrophic bacteria used in the mining and metallurgical process. These cells derive the energy required for their metabolism from the aerobic oxidation of reduced sulfur compounds, including sulfides, elemental sulfur, and thiosulphate, and/or Fe^{+2} can be used as an energy source. The two main functions of this type of bacteria are oxidation of Fe(II) to Fe(III), and S to H_2SO_4, which take part in leaching (Asghari et al., 2013; Lee and Pandey, 2012; Shailesh et al., 2016). Proton-promoted and ligand-promoted mineral solubilization occurs simultaneously in the presence of ligands under acidic conditions (Nancharaiah et al., 2016).

Depending on their tolerance to temperature, the acidophilic micro-organisms are categorized into three groups such as mesophiles, moderate thermoacidophiles, and extreme thermoacidophiles. *Acidithiobacillus ferro-oxidans, Acidithiobacillus thiooxidans*, and *Leptospirillum ferrooxidans* are the important mesophiles which are effective in the temperature range 28–38°C, whereas *Sulfobacillus thermosulfidooxidans*, a moderate thermo-phile can live at 50°C. The *Sulfolobus* species such as *S. acidocaldarius*, *S. solfataricus*, and *S. brierley* are the examples of extreme thermophiles which can be used up to 70°C (Asghari et al., 2013; Lee and Pandey, 2012; Shailesh et al., 2016).

Heterotrophic microorganisms require organic supplements for growth and energy supply. These microorganisms, which include both bacteria and fungi species, are known for their leaching capabilities, especially of oxidic, siliceous, or carbonaceous material. The by-products from the metabolized organic carbon interact with the mineral surface, which causes a partial or complete alteration or dissolution of it. These metabolic by-products are usually organic acids such as acetic, citric, oxalic, tartaric, and keto-gluconic acid, and they play a greater role in mineral dissolution due to their dual effect of lowering the lixiviant pH and metal solubilization by complexing/chelating into soluble organo-metallic complexes. Fungal bioleaching occurs by either of the following mechanisms: (i) acidolysis, (ii) complexolysis, (iii) redox-olysis, and/or (iv) bioaccumulation of metals by the organism's mycelium (Asghari et al., 2013; Bayraktar, 2005; Nancharaiah et al., 2016). First, three processes occur through metabolites excreted by the fungus, and the stability of metal complexes also reduces the toxicity of metal ions to the microorganism by formation of metallo-organic molecules (Kushwah et al., 2015).

In acidolysis, which is also called proton-induced metal solubilization, microbial secretion of protons results in changes of the metal mobility and protons are bound to the surface resulting in the weakening of critical bonds as well as in the replacement of metal ions leaving the solid surface. Heterotrophic fungi like *Aspergillus niger*, *Penicillium simplicissimum*, *Penicillium bilaiae*, *Saccharomyces cerevisiae*, *Yarrowia lipolytica*, etc. have been reported to grow in the presence of electronic scrap and some of them has been used for bioleaching of spent catalysts. These fungi produce organic acids which act as complexing agents and help in extraction of metals like Cu, Cd, Sn, Al, Ni, Pb, Zn, etc. (Shailesh et al., 2016). The effectiveness of fungal bioleaching depends on the production capacity of these extracellular products and their metal resistance. Commercially, fungal bioleaching has proved ineffective due to the cumbersome downstream processing and production of large biomass in addition to low recovery of metals (Asghari et al., 2013; Bayraktar, 2005; Sukla et al., 2014).

Microbial mobilization by heterotrophs cyanogenic bacteria produce cyanide in the aqueous medium forming cyanide complexes from solid materials which has been used to recover silver, gold, and platinum by *Chromobacterium violaceum*, *Pseudomonas fluorescens*, *Pseudomonas plecoglossicida*, and *Bacillus megaterium* (Brandl et al., 2008; Lee and Pandey, 2012). Heterotrophic microorganisms also produce exopolysaccharides, amino acids, and proteins that can solubilize the metals through different mechanisms (Sukla et al., 2014). Several species of actinomycetes (among them *Streptomyces fungicidicus*, *Streptomyces aureofaciens*, and *Streptomyces chibaensis*) have been used to mobilize rare earth elements (REE) from sandy and silty soil samples (Hewedy et al., 2013). In suspensions of 10 g soil per liter (cultivated for 48 h at 30°C), REE leaching efficiencies of up to 37% were obtained depending on the strain applied (Barmettler et al., 2016).

The complex structure of Gram-positive and Gram-negative microorganisms implies that there are many ways for the metal to be taken up by the microbial cell (Abbas et al., 2014). The highly charged nature of lipopolysaccharides confers an overall negative charge on the Gram-negative cell wall. The anionic functional groups present in the peptidoglycan, teichoic acids, and teichuronic acids of Gram-positive bacteria and the peptidoglycan, phospholipids, and lipopolysaccharides of Gram-negative bacteria were the components primarily responsible for the anionic character and metal-binding capability of the cell wall; depending on pH, there may be significant adsorption of heavy metals onto the microbial surface (Barmettler et al., 2016; Vijayaraghavan and Yun, 2008). Extracellular polysaccharides are also capable of binding metals. However, their availability depends on the bacterial species and growth conditions (Abdi and Kazemi, 2015). Naturally occurring microbes can sorb a variety of metals including Pt, Fe, Ni, Cu, Zn, Pb, Cu, Pd, Ag, Cd, Pt, Au, and Hg, with binding capacities typically on the order of 10^{-5}–10^{-3} mol metal g^{-1} (dry wt) microbe. These processes are critically regulated by the chemical groups displayed on the extracellular surfaces of microbial cells, such as carboxyl, phosphoryl, hydroxyl, carbonyl, and amino group (Abbas et al., 2014; Kushwah et al., 2015; Zhuang et al., 2015).

The uptake capacities of bacteria generally range between 0.23 and 0.90 mmol/g. Bacteria species such as *Bacillus*, *Pseudomonas*, *Streptomyces*, *Escherichia*, *Micrococcus*, etc. has been tested for uptake metals (Abbas et al., 2014). In fungal cell walls are complex macromolecular structures mainly consisting of chitins, glucans, mannans, and proteins, and also containing other polysaccharides, lipids, and pigments, e.g., melanin (Abbas et al., 2014). In the fungal cell wall, several types of ionizable sites affect the metal uptake capacity: phosphate groups, carboxyl groups on uranic acids and proteins, and nitrogen-containing ligands on protein as well as on chitin or chitosan (Abbas et al., 2014). Understanding of metal binding groups is an important foundation of a cell surface engineering approaches with the goal of enhancing biosorption efficiencies of heavy metal ions of microbial cells (Cai et al., 2016).

3.3 Growth Medium Composition and Appropriate Level of Nutrients

The limiting factor for bacterial growth may be a lack of inorganic elements such as nitrogen or phosphorus. However, carbon is the factor that limits bacterial growth (Frias et al., 2001), and some microorganisms need nutrients and growth factors to growth in the presence of solid wastes and enhancement of the biodegradation. The microbial cells have in their composition mainly the elements such as C, H, O, N, S, and P. Because of this reason it is necessary to modify the concentration of these macro elements in growth media when the waste does not have a sufficient quantity of them. The waste can be enriched with carbon, nitrogen, phosphorus, and/or sulfur. The microorganisms need macronutrients and micronutrients such as K, Mg, Na, Ca, Fe, and Cr, Co, Cu, Mn, Mo, Ni, Se, W, V, Zn such for their growth and also to carry out enzymatic activities, (Tay et al., 2004). It has been shown that REE are strictly necessary for the growth of some microbes because key enzymes require them as essential cofactors (Barmettler et al., 2016).

During metabolism, microorganisms generate products such as hydrogen, oxygen, and H_2O_2, which can be used for oxidation/reduction of metals. Solubilization or precipitation of metals usually is the result of reduction or oxidation of them. Microbial metabolites can also mediate these two processes. Fermentation and aerobic oxidation carried out by microorganisms can produce organic acids or inorganic acids (nitric and sulfuric acids), which promote the formation of dissolved metal chelates. Phosphate, H_2S, and CO_2, also produced by microorganisms, will stimulate precipitation of non-dissolved carbonates, phosphates, and sulfides of heavy metals such as Cd, As, Cr, Hg, Co, Ni, and Pb; production of H_2S by sulfate-reducing bacteria is especially useful to remove heavy metals and radionuclides from sulfate-containing mining drainage waters. During the anaerobic fermentation of cellulose, organic acids are produced, which can be used as a source of reduced carbon for the reduction of sulfate as well as the additional precipitation of metals (Girma, 2015a,b). Chemolithoautotrophic mesophilic bacteria such as *A. ferrooxidans* and *A. thiooxidans* oxidize ferrous iron to ferric iron and oxidizes sulfur by reducing S^0 to SO_4^{2-} as an energy source and consume CO_2 as a carbon source (Heydarian et al., 2018). The important role of Fe^{2+} may be associated with its dual biological and chemical function. Fe^{2+} is a key substrate for Fe/S oxidizing bacteria and iron is a key factor for the effectiveness of bioleaching process. Several studies have confirmed that Fe^{2+} enhances *A. ferrooxidans* resistance to some metals and the ferric iron produced by bacteria is a strongly oxidizing agent (Beolchini et al., 2010). In studies on the effect of nutrient concentration on metal leaching from spent petroleum catalyst, the Fe(II) concentration was varied from 0 to 9 g/L for iron oxidizing bacteria (IOB) and 0 to 20 g/L for sulfur oxidizing bacteria (SOB). The leaching rates increased as Fe(II) concentration increased up to 3 g/L in IOB but a decrease in leaching rate beyond a certain Fe(II) concentration may be due to a decrease in bacterial activity or due to formation of product layer (precipitated iron) or both. Leaching rate of SOB increased with an increase in elemental sulfur from 10 to 20 g/L in the one step bioleaching media, which may be due to an increase in metabolite concentration (bacterial acid formation) as well as bacterial activities (Barmettler et al., 2016; Mishra et al., 2007; Pradhan et al., 2009).

Bioleaching of REE from spent industrial catalysts and luminescent powder originating from cathode ray tubes (CRT) in the presence and absence of ferric iron was investigated, using heterogenic culture of sulfur oxidizing *A. ferrooxidans*, *A. thiooxidans*, and *L. ferrooxidans*, and results showed that the addition of Fe^{2+} was assumed to enhance bacterial resistance to high levels of metals (Barmettler et al., 2016). Reports of the effect of two different Fe^{+2} concentrations (2 and 32 g/L), added to 9 K medium, on the ability to remove Ni and V contained in a spent catalyst at 16% (w/v) indicated that isolates coded as *Microbacterium liquefaciens* MNSH1-PHGII-1 and *Rhodotorula mucilaginosa* MV-9K-4 showed the maximum Ni and V removal from spent catalyst using Fe^{2+} concentration of 2 g/L in comparison when Fe^{2+} was used at 32 g/L (Gómez-Ramírez et al., 2014b).

In the case of mobilization of REE from mineral solids, the addition of lanthanum and cerium showed an increase of methanol dehydrogenase activity in *Methylobacterium radiotolerans*, *Methylobacterium zatmanii*, and *Methylobacterium fujisawaense* by a factor of 4–6 suggesting the induction of latent genes. Concentrations of lanthanum as low as 2.5 nM stimulated growth on methanol in comparison to the addition of only calcium due to the increased activity of a lanthanide-dependent methanol dehydrogenase. Due to this fact, it has been suggested that methylotrophic microorganisms might find an application in REE biomining and recycling (Barmettler et al., 2016). To cyanogenic bacteria, glycine is a direct precursor of cyanide which is formed as a secondary metabolite by oxidative decarboxylation. In the presence of cyanide, many metals and metalloids (such as Ti, V, Cr, Mn, Fe, Co, Ni, Cu, Zn, Ge, Mo, Tc, Ru, Rh, Pd, Ag, Cd, W, Re, Os, Ir, Pt, Au, Hg, Tl, Po, and U) form well-defined cyanide complexes which show often very good water solubility and exhibit high chemical stability. These findings demonstrate the potential of microbial mobilization of metals such as cyanide complex from solid materials and represent a novel type of microbial metal mobilization termed "biocyanidation" (Brandl et al., 2008). During bioleaching by organic acids, the amount of organic acid produced by some fungi is determined by many factors, which include the buffering capacity of the medium, the carbon source, the ratio of nitrogen and phosphate in the medium, and the experimental conditions for fungal growth. The production of organic acids from sucrose involves a large number of enzymatic processes (Asghari et al., 2013).

The biogenically produced organic acids such as acetic, citric, gluconic, itaconic, lactic, oxalic, and succinic acid played a direct and important role in the bioleaching process and their presence and concentration depended on fungal strain used as well as composition of growth media and carbon source (Asghari et al., 2013; Barmettler et al., 2016). Aung and Ting (2005) observed that an increase in the leaching of the heavy metals is parallel to an increase in the concentration of organic acids. During bioleaching of Ni and V from a power plant residual ash (PPR ash) by organic acids produced by *A. niger*, a decreasing carbon source concentration causes a drop in citric acid production yield and results in accumulation of oxalic acid. The citric acid acts more effective than oxalic acid in leach and recovery of nickel by PPR; however, oxalic acid is also a strong leaching agent in vanadium recovery from PPR (Rasoulnia and Mousavi, 2016).

On the other hand, metal extraction from e-wastes showed that copper in the solution increased with the addition of 1 g/L of citric acid as chelating agent. Shailesh et al., in 2016, carried out a research in which they found that 37%–40% of Cu was leached without adding citric acid and 81%–83% of Cu was leached with the addition of it. The negative effect of the presence of spent catalyst on citric acid production has been also reported; the fungus in the buffered system excreted a marginally higher concentration of oxalate and a lower concentration of gluconate at all pulp densities. The lower concentration of citric acid produced in the buffered culture is mainly due to the maintenance of higher pH (i.e., pH 6), while higher pH results in higher accumulation of oxalic acid due to the induction of the enzyme oxaloacetate hydrolase by the novo synthesis (Asghari et al., 2013). Therefore, the dissolution process is bacterial assisted and depends on the bacterial activity and thus the amount of available nutrients. Goyal et al. (2003) observed that supplementation of fermentation media with cysteine, ammonium sulphate, phosphate and ammonium chloride, and glucose for the growth of *S. cerevisiae* led to different functional groups in the cell surface. Cysteine inserts S- and N-ligands, glucose C-ligands, ammonium N-ligands, and phosphates P-ligands, because all of these have the highest adsorption capacity. Gómez-Ramirez et al. in 2018 observed the effect of glucose on Ni and V removal from a spent catalyst by *Bacillus* spp. strains isolated from mining sites; results showed that the maximum removal values obtained were 2542 mg/kg for Ni and 3701 mg/kg for V in the system, where PHGII medium was not added with glucose, possibly metal interaction capability may be altered by the presence or absence of glucose or distinctive metabolic glucose-dependent and independent, signaling pathways may confer to each strain diverse resistance mechanisms and metal uptake patterns. *Providencia* sp. JAT-1 can grow in the presence of urea and produces ammonia; the ammonia produced by the strain is the main lixiviant agent in bioleaching of low-grade complex copper ore (Hu et al., 2016).

Due to the wide microbial diversity and the different types of metabolisms that can be present in the different microbial groups, one of the important factors during the treatment of solid wastes contaminated with metal is the culture medium used for microbial growth and solid waste treatment because its chemical composition can favor the resistance to metals or increase the synthesis of biomolecules involved in the processes of immobilization or mobilization of metals by the microorganisms.

3.4 Chemical and Metal Composition of Solid Wastes

Hazardous wastes like spent catalyst from petrochemical industry, e-waste, metallurgical slag, mining waste, coal combustion wastes, and other solid wastes are important sources of heavy metals (Gómez-Ramírez et al., 2015, 2018a,b; Rojas-Avelizapa et al., 2018; Zhang et al., 2013). Among the molecular mechanisms that determine the toxicity of heavy metals are: the displacement of essential metallic ions of biomolecules and blocking of their functional groups; the modification of the active conformation of biomolecules, especially enzymes and polynucleotides; the breakdown of the integrity of biomolecules; and the modification of other biologically active agents (Cuizano et al., 2010; Gholami et al., 2011). Knowing the chemical and metallic composition of the waste to be biologically treated will allow selecting the most suitable microorganism to be used in the treatment of solid waste and/or submitting the residue to a previous treatment before being exposed to the microorganism or group of these to improve the efficiency of metal recovery or reduce the metal load of the waste for a safe disposal.

Spent hydroprocessing catalysts are known to contain a variety of potentially toxic metals, and, therefore, studies on the bioavailability and mobility of these metals are critical for understanding the possible environmental risks of the spent catalysts. The presence of metals in particular fractions such as (i) exchangeable fraction, (ii) reducible fraction, (iii) oxidizable fraction, and (iv) residual fraction can influence its susceptibility to bioleaching (Pathak et al., 2018). The recovery of metals from combustion ashes depends on the physical properties and the composition of the ashes (Rojas-Avelizapa et al., 2018; Yao et al., 2014).

Treatment of e-wastes by microbial pathway is limited due to their heterogeneity and their toxicity (Shailesh et al., 2016). Higher concentration of other pollutants could reduce the removal of metals (Abdi and Kazemi, 2015). In addition, an increase in the number of metals decreased bioleaching process due to a number of metallic ions competing for the same binding sites available on the surface of the biomass (Abdullah et al., 2018). So, microbial survival in solid wastes depends on intrinsic biochemical and structural properties, physiological, and/or genetic adaptation including morphological changes of cells and, very importantly, the type and provenance of solid waste used.

3.5 Size of Particle of Solid Wastes

Another factor involved in the treatment of solid wastes is the particle size; few investigations are focused on their influence on metal removal by microorganisms, and one of them was made by Mehta et al. (1999). They carried out studies on bioleaching of Ni, Co, and Cu from copper converter slag using *Thiobacillus ferrooxidans*. The particle sizes used were $+200 \, \mu$m, $-200 + 150 \, \mu$m, $-150 + 100 \, \mu$m, $-100 + 75 \, \mu$m, and $-75 \, \mu$m, observing that the optimum metal dissolution of Ni (50%) and Co (64%) was with particle size of $-75 \, \mu$m. However, in the case of Cu, its recovery decreased from 96% to 66% at the same particle size; they mentioned that it could be due to poor permeation of the leachate to oxidize the copper present in the fine size. Pradhan et al. (2009, 2010), using three different particle sizes from spent petroleum catalyst (45, 106, and 212 μm), observed that leaching rate increased as the particle size decreased for both acidophilic *A. ferrooxidans* (IOB) and *A. thiooxidans* (SOB) strains. This may be due to the increased surface area, which ultimately led to an increase in the depth of penetration of the lixiviant species into the particle, thereby increasing the leaching rate. Lee and Pandey (2012) used *A. ferrooxidans* and *L. ferrooxidans* to recover Mo from waste catalyst of coal liquefaction and observed an increased metal recovery with decreasing particle size of the catalyst. Studies made by Wang et al. (2009) showed that metal leachates from waste printed boards (PWBSs) are influenced by particle size and concentration of the scrap.

When they used *A. ferrooxidans* and/or *A. thiooxidans*, higher metal leaching was observed with the lower size of the particles and decreased pulp density; the percentages of copper, lead, and zinc solubilized were more than 88.9% when fractions were less than 0.35 mm. Bajestani et al. (2014) used *A. ferrooxidans* to bioleach heavy metals from spent Ni–Cd and NiMH household batteries AA of 1.2 V and demonstrated that the highest simultaneous metal extraction of Ni, Cd, and Co were carry out by using a particle size of 62 m.

Santhiya and Ting (2005) made studies during bioleaching metals from spent catalyst using the particle sizes of 100–150 μm, <37 μm, and an average of 2.97 μm, and they observed that increase of oxalic acid secretion by *A. niger* corresponds with a decrease in the catalyst particle size (up to <37 μm), leading to an increase in metal extraction. The highest extraction of metal values from the spent catalyst at 1% w/v pulp density and particle size <37 μm was found to be of Al (54.5%), Ni (58.2%), and Mo (82.3%) at the end of 60 days of bioleaching, and it was observed that the lowest fungal biomass yield occurred in the presence of spent catalyst of an average particle size of 2.97 μm, possibly due to damages in the structure of the fungus caused by the fine size of spent catalyst. Nemati et al. (2000) used batch bioreactor to investigate the effects of particle size on bioleaching of pyrite by using an acidophilic thermophile *Sulfolobus metallicus* and found that decreasing the particle size from 202 to 42.5 μm enhanced the bioleaching rate from 0.05 to 0.098 kg m^{-3} h^{-1}; they also mentioned that a reduction in particle size of the mineral (average diameter of 6.4 μm) adversely influenced the activity of the cells possibly due to damage in the structure of cells as has been reported by Santhiya and Ting in 2005, resulting in their inability to oxidize pyrite.

There are several reports on the biological treatment of solid wastes where the particle sizes used were varied depending on the waste to be treated, being generally between 3 and 1000 μm, generally the used size (Barmettler et al., 2016; Gómez-Ramírez et al., 2014a, 2018a; Lee and Pandey, 2012; Mehta et al., 1999; Nemati et al., 2000; Pradhan et al., 2009; Santhiya and Ting, 2005; Wang et al., 2009). To achieve a greater recovery of metals, the size of the particle is not the most important factor, also affect the concentration of the waste to be used, temperature, pH, contact time, nature of the waste, etc. These investigations indicate that particles of very small sizes can affect at several levels within which we can include the effect on the growth of microorganisms due to the damage of microbial cellular structures. On the other hand, these particles of small sizes can favor the extraction of some metals and with their decrease in size, the production of some important metabolites for the mobilization or immobilization of metals by the microbial route can be improved. However, a small particle size does not always improve the metal recovery process.

3.6 Pulp Density

In biohydrometallurgy/bioleaching, pulp density, i.e., the ratio of solid materials weight (w or s) to volume of bioleaching media (v) (Niu et al., 2015; Xin et al., 2012) is one of the most important variables which can affect the bioprocess of metal removal and determines the commercial applications of bioleaching of different solid wastes. It affects the microbial adhesion on solid waste, and oxygen mass transfer from gas to liquid phase can increase the release of toxic chemicals or electrolytes from solids to the solution (Ferreira et al., 2017; Gómez-Ramírez et al., 2014a; Niu et al., 2014; Pradhan et al., 2009; Qu and Lian, 2013; Suzuki et al., 1999; Wang et al., 2009). The above causes lower growth and reduction or inhibition of important metabolites involved in mobilization or immobilization of metals by autotrophic and/or heterotrophic microorganisms.

Generally, during biotreatment of solid wastes containing metals, a low pulp density is used with ranges between 0.01% and 10% (w/v) and rarely higher than 16% (w/v) to 80% (w/v) (Arenas-Isaac et al., 2017; Bayraktar, 2005; Bharadwaj and Ting, 2013; Gómez-Ramírez et al., 2015, 2014a,b, 2018b; Mishra et al., 2007; Pradhan et al., 2009; Rivas-Castillo et al., 2018; Rojas-Avelizapa et al., 2015). In biohydrometallurgy of low-grade ores, pulp density is generally 10% or higher because the ores are mainly reduced sulfides ores which do not contain alkaline matter or toxic compounds. In contrast, the spent batteries and spent catalyst could contain high concentration of alkaline matter or toxic compounds as oxides or hydroxides which greatly harm growth and activity of leaching cells. As a result, the pulp density was often only 1% or lower in bioleaching of the spent batteries and spent catalyst (Niu et al., 2015). Few reports are focused on their effect on metal removal as will be mentioned below. Tipre and Shailesh (2004) carried out bioleaching of metals from a Cu–Pb–Zn sulfide bulk concentrate at different pulp densities of 5%–25% (w/v) using a developed consortium by *A. ferrooxidans, A. thiooxidans, L. ferrooxidans*, and heterotrophic organisms. In this study, an increase in pulp density up to 20% (w/v) increased the extraction of Cu and Zn, but at 25% (w/v), a decreased rate of metal extraction was observed. It was also found that continuous bioleaching with controlled addition of pulp could provide an economically viable technology for metal extraction from polymetallic bulk concentrate.

Mishra et al. (2007) used a sulfur oxidizing lithotrophic bacteria to bioleach V from spent refinery; different pulp densities of 5–50 g/L were used. After 16 days of incubation, Ni recovery decreased from 89.0% to

76.9% at 5 and 50 g/L of solid, respectively, although this was not significant. However, with increased catalyst concentration, V extraction fell from 89.1% to 16.4% and Mo recovery fell from 78.6% to 42.6%; so they conclude that metal dissolution behavior in the cultures with increasing spent catalysts concentration remains unclear. Pradhan et al. (2009–2010) conducted studies of bioleaching by an iron oxidizing (IOB) and sulfur oxidizing bacteria (SOB) at different pulp densities from 5% to 25% of spent petrochemical catalyst. In the case of IOB, the leaching rate decreased as the pulp density increased, and as the pulp density increased, there was a gradual depletion of oxygen, thereby making the bacteria less active, which resulted in a low leaching rate. When SOB were used, leaching rate of some metals such as Ni and V increased, as did the pulp density, while Mo leaching rate decreased when the pulp density was greater than 10%.

Xu and Ting (2009) used various pulp densities of 1%–5% in a batch system during bioleaching metals by *A. niger* from fly ash, finding that an increase in the pulp density led to a decrease leaching yield in order of 97% of Al and 98% of Zn at 1% pulp density and at 5% pulp density, the extraction yield decreased to 60% and 57% to Al and Zn, respectively. For iron, the extraction yield was 56% and 48%, respectively, at 1% and 5% of pulp density, and no growth was observed at 6% of it, possibly due to conformational changes of polymers in the cells, the blocking of essential functional groups of enzymes, the displacement of essential metals, and modifications in membrane integrity as well as transport processes.

Amiri et al. (2011) examined the growth kinetics of *P. simplicissimum* in the presence of spent catalyst at different pulp densities of 2%–11% w/v, finding that maximum extraction yields were observed in a two-step bioleaching process at 3% w/v. They also found that in two-step fungal leaching, fungal germination takes place before addition of solid to the medium and, in this way, limiting the inhibition effect of the spent catalyst on fungal growth and organic acid production. On the other hand, the excretion of organic acids before addition of spent catalyst caused a decrease in the toxicity of heavy metal ions through complexation (Asghari et al., 2013).

Qu and Lian (2013) carried out bioleaching of rare earth and radioactive elements from red mud by *Penicillium tricolor* and observed that increasing pulp density to 2–10 g/L, REE leaching decreases. It was also observed that under one-step bioleaching process, the higher efficiencies were at 2% of pulp density. However, the highest extraction yields were achieved under two-step process at 10% (w/v) pulp density, and the main lixiviants in the bioleaching process were oxalic and citric acids.

Niu et al. (2014) used a mixed culture of sulfur-oxidizing bacteria *Alicyclobacillus* sp. and the iron-oxidizing bacteria *Sulfobacillus* sp. to release Co and Li from spent lithium ion batteries (LIBs) by bioleaching; three pulp densities of 1%, 2%, and 4% were used, noting that by increasing pulp density from 1% to 4%, the bioleaching efficiency decreased from 52% to 10% for Co and from 80% to 37% for Li after an incubation period of 11 days, and they mentioned that at higher pulp density, high amount of toxic electrolyte was dissolved into bioleaching media, endangering the growth and activity of both kinds of bacteria in spite of spent LIBs tolerance.

Gómez-Ramirez et al. in (2014a) demonstrated the sulfur removal of spent catalyst by *Acidithiobacillus ferrooxidans* ATCC 53987 and *A. thiooxidans* AZCT-M125-5 at different pulp densities of 8.25%, 16.5%, 25%, 33%, 49.5%, 66.2%, 82.2%, and 100% w/v, corresponding to sulfur contents of 0.7%, 1.4%, 2.1%, 2.8%, 4.3%, 5.7%, 7.1%, and 8.63% (w/w), respectively, showed that pulp densities higher than 33% and 49.5% (w/v) totally inhibited the sulfur-oxidizing activity of *A. thiooxidans* and *A. ferrooxidans*, respectively. Such results may be attributed to the lower water content in the medium which causes reduction of mass transport and accessibility of microorganisms to elemental sulfur, thus compromising their metabolic activities. It is well established that folded conformation of proteins in solution is determined largely by water properties; cells lose water when exposed to high osmotic solutions, and such conditions might be deleterious to the maintenance of their normal structure, function, and metabolic control (Suzuki et al., 1999).

Hu et al. in 2016, after adaptation of heterotrophic bacterium *Providencia* sp. JAT-1 to copper ore, carried out bioleaching experiments at different pulp densities of 1%, 3%, 5%, 7%, 9%, and 11% of low-grade copper ore, showing that copper extraction decreased with the increase of pulp density and direct relationship between copper extraction and growth was observed. The copper extraction reached the highest value of 54.5% at the pulp density of 1% and the ammonia concentration involved in bioleaching decreased with the increase of pulp density.

Ferreira et al. (2017) used *A. thiooxidans* FG-01 to bioleach metals from spent diesel hydrodesulfurization catalyst, with different concentrations of spent catalyst of 5–50 g/L during seven days, observing that sulfuric acid production was affected in all concentrations used; so the bioleaching process was conducted at 5 g/L of pulp density, and they concluded that the solubilized metal concentration and the presence of the other contaminants had an important influence on the cell activity and consequently the bioleaching process. In 2018, Heydarian et al. carried out bioleaching of metals from spent lithium-ion batteries (LIBs) by using a mixture of *A. ferrooxidans* and *A. thiooxidans*, where previously, during 128 days, the bacteria were adapted starting from 2.5 g/L (0.25%) of pulp density and ending at 40 g/L (4%) of pulp density. The bacteria became inactive at pulp density above 40 g/L. The maximum recovery of metal at 40 g/L was about 99.2% for Li, 50.4% for Co, and 89.4% for Ni under optimum conditions of 36.7 g/L of iron sulfate, 5.0 g/L of sulfur, and initial pH of 1.5.

All these data show us the preference of using low pulp densities for the treatment of solid wastes contaminated with metals, a pre-treatment of waste before being exposed to microorganisms, or sequential processes using one or more microorganism and, in some cases, the production of metabolites involved in the processes of mobilization or immobilization of metals previously without the residue to reduce toxicity problems, and thus increasing densities of the pulp to be biotreated. However, more studies are needed to understand the interaction of the microorganisms with solid wastes at high concentrations and thus improve biological treatments.

3.7 pH

The pH of environmental sites varies from 1 to 11 depending on site; pH values of 1–3 are found in volcanic soil and mine drainage, etc.; acid soils have pH values between 3 and 5; fresh and sea water have pH between 7 and 8; alkaline soils, lakes, and solutions of ammonia have pH values between 9 and 11 (Tay et al., 2004). Based on their pH optimum of growth, microorganisms are grouped in acidophilic, neutrophilic, and alkalophilic. Species that grow at pH lower than 4 are called acidophilic, species adapted to grow at pH values higher than 9 are called alkalophilic, and species whose pH range of growth is 5.5–8 are called neutrophilic (Abatenh et al., 2017; Mishra and Rhee, 2010); pH has impacted on microbial metabolic activity and can also increase or decrease removal process of metals.

Changes of pH during biotreatment can show different results because metabolic processes are highly susceptible to even slight changes in pH (Abatenh et al., 2017). Therefore, pH of the medium of growth during treatment of solid wastes or bioremediation is important to maintain at optimal biotreatment because some wastes can cause acidification/alkalinization during biotreatment process depending on their chemical composition or pulp density used. Krebs et al. (1997) mentioned that during biohydrometallurgical processing of fly ash at high pulp densities, high content of toxic metals causes serious problems, also the salinity and strong pH (greater than 10) that originates in the environment.

On the other hand, the presence of solid wastes at 80% pulp density containing higher metal concentrations can reduce pH of growth media from 7 to 2.8 units, probably due to metals which were partially solubilized causing a pH decrease (Rojas-Avelizapa et al., 2015). Kumar and Nagendra (2007) investigated the influence of initial pH on bioleaching of heavy metals from contaminated soil by indigenous *A. thiooxidans* using different initial pH of 3–7. The solubilization of Cd, Cr, Co, Pb, and Zn was maximum at initial pH of 5, 6, and 7. They suggest that initial acidification of the bioleaching system is not a prerequisite for growth and acid production by *A. thiooxidans*. Generally, the pH during treatment can be changed due to organic acids produced during fermentation or by formation of inorganic acids formed in aerobic oxidation of ammonium, elemental sulfur, hydrogen sulfide, or metal sulfides, which leads to a decrease in pH. The pH increase can be attributed to the initial dissolution of metallic hydroxides and oxidation of ferrous ions which take protons. The pH value can influence on: the microbial metabolic activity, the carbon availability, nutrient availability, the solubility of metals, speciation, modification of the solution chemistry of the heavy metals, the activity of the functional group in the biomass and the competition of the metallic ions, and increase and decrease of removal process of metals can occur (Abatenh et al., 2017; Bajestani et al., 2014; Das et al., 2008; Pradhan et al., 2010; Rousk et al., 2009; Wang and Chen, 2006; Xu and Ting, 2009).

Some microorganisms belonging to acidophilic group help to dissolve metals from solid phase of wastes into the aqueous phase; so the pH of the system is known to be the most important operating parameter that influences metal solubilization during bioleaching and plays an important role in the activity of acidophilic bacteria. Among the bacteria *A. ferrooxidans*, *A. thiooxidans*, *L. ferrooxidans*, and *Sulfolobus* sp. are well-known consortia for the bioleaching activity, while *Penicillium* and *A. niger* are some fungi that help in metal leaching process (Kumar and Nagendran, 2007; Mishra and Rhee, 2010; Pradhan et al., 2010; Xu and Ting, 2009). The production of organic acids provides anions and protons that promote leaching by acidolysis and complexolysis. In acidolysis, the metal is solubilized through organic or inorganic acid produced by microorganisms. In complexolysis, the anions form complexes with the metal cations to produce metal complexes. Low pH (1–2) favors acidolysis, with the release and enhanced mobility of free metal cations by protonation and increasing the bioleaching (Kumar and Nagendran, 2007; Pradhan et al., 2009; Xu and Ting, 2009).

Autotrophic organisms have a high requirement for compounds such as NAD(P)H to reduce their carbon source (CO_2) to produce nucleotides, sugars, amino acids, and other molecules important to synthesized new cell mass. Chemolithotrophic autotrophs require a large transmembrane proton gradient to generate the required proton motive force to energize the synthesis of NAD(P)H. The process is known as reverse electron transport. The growth in acid solutions is a nutritional necessity as a large transmembrane pH gradient is required to produce the hydrogen atoms needed to reduce CO_2 to cell mass. In the case of heterotrophic bacteria, they do not have a high demand for NAD(P)H as their carbon source is more reduced than CO_2, and hydrogen atoms removed from their source of nutrition may be used to satisfy their lower NAD(P)H requirement (Mishra and Rhee, 2010). During biotreatment of solid waste by iron-oxidizing (IOB) and sulfur-oxidizing (SOB) bacteria, generally, a lower pH of 1–3 is used (Kumar and Nagendran, 2007; Pradhan et al., 2009, 2010; Xu and Ting, 2009).

Studies made by Pradhan et al. (2009) evaluated the initial effect of pH 1.5–3 on metal bioleaching from spent catalyst. Results showed that leaching rate for Ni, V, and Mo increase with an increase in pH up to 2.5 for both species and a further increase in pH reversed this trend. With respect to IOB and SOB, the iron and sulfur oxidation rates were maximal at pH 2.5 and 2.0, respectively. At pH values greater than 2.5, a slight precipitation of iron was observed. Therefore, the decrease in leaching rate at pH values greater than 2.5 was mostly likely due to a decrease in bacterial activity as well as the formation of a product layer such as precipitated iron.

Abdullah et al. in 2018 carried out studies about the removal of heavy metals from MSW, observing that at pH 5.5 removal was better than at pH of 4. The improved performance at pH 5.5 could be correlated with higher metal solubility, generally for different biosorption systems for metal ions. Optimal pH differs depending on the metal; in this study, the optimal pH for biosorption by *S. cerevisiae* is of 5–9 for Cu and 4–5 for uranium. In general, the optimum bioleaching pH was above pH 5. For biosorption of heavy metal ions, pH is one of the most important environmental factors. The biosorption capacity of metal cations increases with increasing pH of the sorption system but not in a linear relationship due to more ligands as phosphate, carbonyl, imidazole, and amino groups, which would be exposed and carry negative charges, and these can attract metals and be bioabsorbed on the cell surface.

On the other hand, too high pH value could cause precipitation of metal complexes due to solubility of metals decreased by formation of M(OH) (Abbas et al., 2014; Abdi and Kazemi, 2015; Wang and Chen, 2006). The activity of binding sites can also be changed by adjustment of the pH value, for example, for the biosorption of metal ions by bacterial biomass, pH 3–6 has been found favorable. A protonated bacterial biomass releases H^+ ions during the biosorption of metals, which in turn decreases the solution pH (Abbas et al., 2014). The affinity of cationic species for the functional groups contained on the cellular surface is strongly dependent on pH. At low pH values, cell wall ligands are closely associated with hydronium ions and restrict the biosorption of metal ion (M^{2+}) as a result of competition between H_3O^+ and M^{2+} with the bacterial cell wall ligands (Abdi and Kazemi, 2015; Sag and Kutsal, 2001). Finally, the use of cyanogenic bacteria allows the mobilization of metals from solids under alkaline conditions and metals might be recovered more easily. Leaching under alkaline conditions (pH 8.4) using ammonifying and alkalophilic chemolithotrophic bacteria has been reported rarely and leachates could be less corrosive at alkaline pH (Brandl et al., 2008; Ibragimova et al., 2007). However, the pH used is related with the microorganism and solid waste to be biotreated.

3.8 Temperature

Among the physical factors, temperature is one of the most important to determining the survival and growth rate of microorganisms in different environments because the rate of microbial activities increases with temperature and reaches its maximum level at an optimum temperature. In this process, the enzymes are involved and each one has an optimum temperature of activity which can decrease suddenly with further increase or decrease in temperature, so variations in temperature also speed up or slow down the biotreatment process because microbial physiological characteristics are highly influenced by it (Abatenh et al., 2017).

Based on optimal temperature growth, the microorganisms are classified as psychrophiles, mesophiles, and thermophiles. Mesophiles microorganisms grow at room temperature (20–40°C). Among the mesophiles involved in metal removal, *A. ferrooxidans*, *A. thiooxidans*, *Acidithiobacillus caldus*, *L. ferrooxidans*, *Leptospirillum ferrodiazotrophum*, *Leptospirillum thermoferrooxidans*, *Leptospirillum ferriphilum*, *Bacillus* sp., *Microbacterium*, *Aspergillus*, and *Penicillium* can be mentioned. Moderate thermophiles are able to grow at a temperature of around (40–60°C). Important moderate thermophiles are *Acidimicrobium ferrooxidans*, *A. caldus*, and *Sulphobacillus thermosulphooxidans* which have the ability to oxidize sulfur and iron. Extreme thermophiles are those which can grow actively even at a temperature of 60–80°C. The most important extreme thermophiles are *S. metallicus, S. acidocaldarius, S. olfataricus, S. brierley, S. ambioalous, Sulphobacillus* sp. and *Metallosphaera sedula*. The oxidation kinetics of the extreme thermophiles is greater than the mesophiles and the moderate thermophiles because these microorganisms can grow at higher temperatures. The bioleaching kinetics carried out by extreme thermophiles is higher than that of mesophiles and moderate thermophiles.

The bioleaching dissolution reaction is exothermic; therefore, the temperature increases during the reaction. So if extreme thermophiles are used, then heat exchanger may not be required to control the leaching temperature. (Barmettler et al., 2016; Mishra and Rhee, 2010; Sukla et al., 2014). In all microorganisms, their affinity to substrates could be decreased consistently as temperature drops below the optimum temperature of growth. This effect may be because of stiffening of the lipids of the membrane below the temperature optimum, leading to decreased efficiency of transport proteins embedded in the membrane. The lower temperature limit for growth is, therefore, that temperature at which an organism is no longer able to supply the maintenance requirement of the growth-rate-limiting nutrient because of loss of affinity for that substrate (Nedwell, 1999). The temperature has a strong influence on the metal removal contained in solid wastes, but the temperature employed will depend, to a large extent, on the type of microorganism and must be maintained at optimal growth temperatures for effective biotreatment. Few investigations are focused on the effect of temperature on metal removal by mobilization or immobilization and are mentioned below.

Mishra et al. (2008) used three different temperatures (25, 32, and 40°C) for bioleaching of Ni, V, and Mo from spent hydroprocessing catalyst by using acidophilic bacteria. Under these conditions, the percentage of Ni extraction varied from 80% to 90% while V from 89% to 97% when increasing temperature. However, Mo dissolution decreased when temperature increase, under these conditions, the leaching percentage of Mo varied from 60% to 17%. Pradhan et al. (2010) studied the effect of temperature on leaching of Mo, V, and Ni by IOB and SOB at temperatures ranging from 10 to 35°C. In both cases, the leaching rate of metal ions increased in the same way as the temperature did. Since bacteria were mesophilic, temperature of 35°C was detrimental for bacterial growth and reduced the metal extraction efficiency.

In the case of bioaccumulation/biosorption process, an increase of temperature usually enhances biosorption by increasing surface activity and kinetic energy of the adsorbate and remains unaffected within the range

20–35°C. In some cases, high temperatures may cause permanent damage to microbial living cells or damage the active binding sites on biomass, decreasing metal uptake. But nevertheless, adsorption reactions are generally exothermic and the adsorption increases with decreasing temperature (Abbas et al., 2014; Abdi and Kazemi, 2015; Wang and Chen, 2006).

Niu et al. (2014) observed that temperature had a strong influence on the bioleaching performance of spent LIBs (spent lithium ion batteries) by *Alicyclobacillus* sp. (SOB) and *Sulfobacillus* sp. (IOB) after acclimation of 30 days by contact with the spent LIB powder at dose from 0.25% to 1.0% (w/v) of pulp density. When temperature increases from 30 to 35°C, the efficiency also increases from 78% to 89% for Li and from 52% to 72% for Co, respectively. However, higher temperature of 40°C adversely affected the activity and growth of the leaching cells which is possibly attributed to its mesophilic nature. The results demonstrated that temperature of 35°C was the optimum for the bioleaching of spent LIBs. As it has been mentioned throughout this chapter, the recovery of metals through the biological route is multifactorial since several factors are involved.

3.9 Inocula Size

The population density used for the treatment of solid wastes contaminated with metals is also an important factor since low inoculum sizes may lead to the establishment of a prolonged lag phase (adaptation phase), requiring a longer time for solid waste biotreatment. However, large sizes of inoculum do not guarantee the efficiency of the process of metal removal. Very few studies are focused on the effect of inoculum size on the treatment of contaminated solid wastes. However, it is mentioned that the biomass concentration in the solution seems to influence the specific uptake of metals and affect biosorption efficiency with a reduction in sorption per unit weight occurring with increasing biomass concentration due to restrictions in the access of metal ions to the binding sites (Abbas et al., 2014; Das et al., 2008).

Generally, the inoculum sizes used for treatment of solid waste by microorganism are fixed in range of 1×10^7, 2×10^7, 3×10^8, 1×10^8, and 6.1–7.8×10^9 cells/mL (Gholami et al., 2011; Gómez-Ramírez et al., 2014a, 2018a,b; Hu et al., 2016; Mishra et al., 2007; Rojas-Avelizapa et al., 2015, 2018). Studies on metal removal by *A. ferrooxidans* through cells and extracellular culture supernatant were carry out by specific metal removal rate (SMRR) using different inoculum sizes. The study explored an innovative metal mechanistic process by *A. ferrooxidans* to remove metal; first, the oxidation of Fe^{2+} to Fe^{3+} was carried out by the bacterium and then the produced Fe^{3+} oxidized the metals. Results show that the optimal cell concentrations used for metal removal were 5.0×10^8 cells/mL at 48 h, 2.0×10^8 cells/mL at 49 h, and 1.0×10^8 cells/mL at 41.5 h for Co, Ni, and Al, respectively. The SMRR achieved were 2.0, 1.6, and 0.55 mg/h cm^2 for Co, Ni, and Al, respectively (Hocheng et al., 2012). Based on the above, the size of inoculum can favor the removal of a particular metal; however, more studies are needed to understand the relationship between inoculum size and metal removal.

3.10 Time Contact and Oxygen Concentration

Different mechanisms are involved in the process of metal removal by microorganisms. Contact leaching describes the direct physical contact between microorganisms and a solid, whereas in non-contact leaching, the biomass is physically separated from the solid to be treated. The time of treatment of solid wastes can change based on different factors such as pulp density, particle size, type of microorganisms, and microbial growth system used (airlift reactor, columns, batch culture, etc.); generally, the period used can be days or months (5 until 180 days) for their treatment (Barmettler et al., 2016; Kim et al., 2018; Lee and Pandey, 2012).

During treatment of solid wastes, different microorganisms could be used; some of them require oxygen and others do not require it for growth. Based on the relation of microorganisms to oxygen, these are classified in anaerobic (living without oxygen), facultative anaerobic (living under anaerobic or aerobic conditions), microaerophilic (preferring to live under low concentrations of dissolved oxygen), and obligate aerobic (living only in the presence of oxygen) microorganisms. Some anaerobic microorganisms, called tolerant anaerobes, have mechanisms protecting them from exposure to oxygen. Others, called obligate anaerobes, may be killed after exposure to aerobic conditions. Obligate anaerobes produce energy from: fermentation, anaerobic respiration using electron acceptors such as CO_2, NO_3^-, NO_2^-, Fe^{3+}, and SO_4^{2+}. Facultative anaerobes can produce energy from these reactions or from the aerobic oxidation of organic matter. Biological treatment of solid wastes is carried out in aerobic or anaerobic conditions.

The majority of the microorganisms used in the treatment of solid wastes contaminated with metals and which were mentioned in this subject, based on their oxygen requirements, fall into the group of aerobes, which is why agitation or oxygen injection conditions are needed to promote the growth and removal of metals. The presence of oxygen in most cases can enhance the metabolism and metal removal by mechanisms of mobilization and immobilization (Abatenh et al., 2017). Also, anaerobic microorganisms have been applied for the mobilization of REE from solid materials. Yttrium was mobilized from phosphogypsum (a by-product originating from fertilizer production) in a fixed-bed reactor by sulfate-reducing *Desulfovibrio desulfuricans* with efficiencies of almost 80% (Barmettler et al., 2016).

Based on the foregoing, the treatment and/or recovery of metals mediated by microorganisms is influenced by several factors such as those explained in this chapter, where each of them plays a very important and crucial role in the success of the biotreatment of solid wastes contaminated by metals.

References

Abatenh, E., Gizaw, B., Tsegaye, Z., Wassie, M., 2017. Application of microorganisms in bioremediation-review. J. Environ. Microbiol, 1(1), 1–9.

Abbas, A.H., Ismail, I.M., Mostafa, T.M., Sulaymon A.H., 2014. Biosorption of heavy metals: A review. J. Chem. Sci. Technol. 3(4), 74–102.

Abdi, O., Kazemi, M., 2015. A review study of biosorption of heavy metals and comparison between different biosorbents. J. Mater. Environ. Sci. 6(5), 1386–1399.

Abdullah, J.J., Emam, A.A.E., Greetham, D., Du, C., Tucker, G.A., 2018. Optimized conditions for bioleaching using yeasts from municipal solid wastes to produce safe compost or fertiliser. Int. J. Environ. Sci. Nat. Res. 8(3), 555738. https://doi.org/10.19080/IJESNR.2018.08.555738

Adebayo, F.O., Obiekezie, S.O., 2018. Microorganisms in waste management. Res. J. Sci. Technol. 10, 28. https://doi.org/10.5958/2349-2988.2018.00005.0

Amiri, F., Mousavi, S.M., Yaghmaei, S., 2011. Enhancement of bioleaching of a spent Ni/Mo hydroprocessing catalyst by *Penicillium simplicissimum*. Sep. Purif. Technol. 80, 566–576. https://doi.org/10.1016/j.seppur.2011.06.012

Arenas-Isaac, G., Gómez-Ramírez, M., Montero-Álvarez, L.A., Tobón-Avilés, A., Fierros Romero, G., Rojas-Avelizapa, N.G., 2017. Novel microorganisms for the treatment of Ni and V as spent catalysts. Indian J. Biotechnol. 16, 370–379.

Asghari, I., Mousavi, S.M., Amiri, F., Tavassoli, S., 2013. Bioleaching of spent refinery catalysts: A review. J. Ind. Eng. Chem. 19, 1069–1081. https://doi.org/10.1016/j.jiec.2012.12.005

Aung, K.M.M., Ting, Y.-P., 2005. Bioleaching of spent fluid catalytic cracking catalyst using *Aspergillus niger*. J. Biotechnol. 116, 159–170. https://doi.org/10.1016/j.jbiotec.2004.10.008

Bajestani, I., Mousavi, S.M., Shojaosadati, S.A., 2014. Bioleaching of heavy metals from spent household batteries using *Acidithiobacillus ferrooxidans*: Statistical evaluation and optimization. Sep. Purif. Technol. 132, 309–316. https://doi.org/10.1016/j.seppur.2014.05.023

Barmettler, F., Castelberg, C., Fabbri, C., Brandl, H., Working Group of Environmental Microbiology, Department of Evolutionary Biology and Environmental Studies, University of Zurich, Winterthurerstrasse 190, CH-8057 Zurich, Switzerland, 2016. Microbial mobilization of rare earth elements (REE) from mineral solids—A mini review. AIMS Microbiol. 2, 190–204. https://doi.org/10.3934/microbiol.2016.2.190

Bayraktar, O., 2005. Bioleaching of nickel from equilibrium fluid catalytic cracking catalysts. World J. Microbiol. Biotechnol. 21, 661–665. https://doi.org/10.1007/s11274-004-3573-6

Beolchini, F., Fonti, V., Ferella, F., Vegliò, F., 2010. Metal recovery from spent refinery catalysts by means of biotechnological strategies. J. Hazard. Mater. 178, 529–534. https://doi.org/10.1016/j.jhazmat.2010.01.114

Bharadwaj, A., Ting, Y.-P., 2013. Bioleaching of spent hydrotreating catalyst by acidophilic thermophile *Acidianus brierleyi*: Leaching mechanism and effect of decoking. Bioresour. Technol. 130, 673–680. https://doi.org/10.1016/j.biortech.2012.12.047

Brandl, H., Faramarzi, M.A., 2006. Microbe-metal-interactions for the biotechnological treatment of metal-containing solid waste. China Particuology 4, 93–97. https://doi.org/10.1016/S1672-2515(7)60244-9

Brandl, H., Lehmann, S., Faramarzi, M.A., Martinelli, D., 2008. Biomobilization of silver, gold, and platinum from solid waste materials by HCN-forming microorganisms. Hydrometallurgy, 17th International Biohydrometallurgy Symposium, IBS 2007, Frankfurt a.M., Germany, 2–5 September 2007 94, 14–17. https://doi.org/10.1016/j.hydromet.2008.05.016

Cai, C.-X., Xu, J., Deng, N.-F., Dong, X.-W., Tang, H., Liang, Y., Fan, X.-W., Li, Y.-Z., 2016. A novel approach of utilization of the fungal conidia biomass to remove heavy metals from the aqueous solution through immobilization. Sci. Rep. 6, 36546. https://doi.org/10.1038/srep36546

Cuizano, N.A., Reyes, U.F., Dominguez, S., Llanos, B.P., Navarro, A.E., 2010. Relevancia del pH en la adsorción de iones metálicos mediante algas pardas. Rev. Soc. Quím. Perú. 76, 123–130.

Das, N., Vimala, R., Karthika, P., 2008. Biosorption of heavy metals – An overview. Indian J. Biotechnol. 7(2), 159–169.

Dzionek, A., Wojcieszyńska, D., Guzik, U., 2016. Natural carriers in bioremediation: A review. Electron. J. Biotechnol. 23, 28–36. https://doi.org/10.1016/j.ejbt.2016.07.003

Ferreira, P.F., Sérvulo, E.F.C., Costa, A.C.A. da, Ferreira, D.M., Godoy, M.L.D.P., Oliveira, F.J.S., Ferreira, P.F., Sérvulo, E.F.C., Costa, A.C.A. da, Ferreira, D.M., Godoy, M.L.D.P., Oliveira, F.J.S., 2017. Bioleaching of metals from a spent diesel hydrodesulfurization catalyst employing *Acidithiobacillus thiooxidans* FG-01. Braz. J. Chem. Eng. 34, 119–129. https://doi.org/10.1590/0104-6632.20170341s20150208

Frias, J., Ribas, F., Lucena, F., 2001. Effects of different nutrients on bacterial growth in a pilot distribution system. Antonie Van Leeuwenhoek 80, 129–138. https://doi.org/10.1023/A:1012229503589

Gadd, G.M., 2010. Metals, minerals and microbes: Geomicrobiology and bioremediation. Microbiol. Read. Engl. 156, 609–643. https://doi.org/10.1099/mic.0.037143-0

Gholami, R.M., Borghei, S.M., Mousavi, S.M., 2011. Bacterial leaching of a spent Mo–Co–Ni refinery catalyst using *Acidithiobacillus ferrooxidans* and *Acidithiobacillus thiooxidans*. Hydrometallurgy 106, 26–31. https://doi.org/10.1016/j.hydromet.2010.11.011

Girma, G., 2015a. Microbial bioremediation of some heavy metals in soils: An updated review. J. Resour. Dev. Manag. 10, 62-73–73.

Girma, G., 2015b. Microbial bioremediation of some heavy metals in soils: An updated review. Egypt. Acad. J. Biol. Sci. G Microbiol. 7, 29–45. https://doi.org/10.21608/eajbsg.2015.16483

Gómez-Ramírez, M., Flores-Martínez, Y.A., Lopéz-Hernández, L.J., Rojas-Avelizapa, N.G., 2014b. Effect of Fe^{2+} concentration on microbial removal of Ni and V from spent catalyst. J. Chem. Biol. Phys. Sci. Spec. Issue Sect. B 4(5), 101–109.

Gómez-Ramírez, M., Montero-Álvarez, L.A., Tobón-Avilés, A., Fierros-Romero, G., Rojas-Avelizapa, N.G., 2015. *Microbacterium oxydans* and *Microbacterium liquefaciens*: A biological alternative for the treatment of Ni-V-containing wastes. J. Environ. Sci.

Health Part A Tox. Hazard. Subst. Environ. Eng. 50, 602–610. https://doi.org/10.1080/10934529.2015.994953

Gómez-Ramírez, M., Rivas-Castillo, A., Rodríguez-Pozos, I., Avalos-Zuñiga, R.A., Rojas-Avelizapa, N.G., 2018b. Feasibility Study of Mine Tailing's Treatment by *Acidithiobacillus thiooxidans* DSM 26636. Int. J. Biotechnol. Bioeng. 12, 4. https://doi.org/10.5281/zenodo.2363155.

Gómez-Ramírez, M., Rivas-Castillo, A.M., Monroy-Oropeza, S.G., Escorcia-Gómez, A., Rojas-Avelizapa, N.G., 2018a. Effect of Glucose Concentration on Ni and V Removal from a Spent Catalyst by *Bacillus* spp. Strains Isolated from Mining Sites. Acta Univ. 28, 1–8. doi:10.15174/au.2018.1475

Gómez-Ramírez, M., Zarco-Tovar, K., Aburto, J., de León, R.G., Rojas-Avelizapa, N.G., 2014a. Microbial treatment of sulfur-contaminated industrial wastes. J. Environ. Sci. Health Part A Tox. Hazard. Subst. Environ. Eng. 49, 228–232. https://doi.org/10.1080/10934529.2013.838926

Goyal, N., Jain, S.C., Banerjee, U.C., 2003. Comparative studies on the microbial adsorption of heavy metals. Adv. Environ. Res. 7, 311–319. https://doi.org/10.1016/S1093-0191(02)00004-7

Hewedy, M.A., Rushdy, A.A., Kamal, N.M., 2013. Bioleaching of rare earth elements and uranium from sinai soil, Egypt using actinomycetes. Egypt. J. Hosp. Med. 53, 909–917. https://doi.org/10.12816/0001653

Heydarian, A., Mousavi, S.M., Vakilchap, F., Baniasadi, M., 2018. Application of a mixed culture of adapted acidophilic bacteria in two-step bioleaching of spent lithium-ion laptop batteries. J. Power Sources 378, 19–30. https://doi.org/10.1016/j.jpowsour.2017.12.009

Hocheng, H., Chang, J.H., Hsu, H.S., Han, H.J., Chang, Y.L., Jadhav, U.U., 2012. Metal removal by *Acidithiobacillus ferrooxidans* through cells and extra-cellular culture supernatant in biomachining. CIRP J. Manuf. Sci. Technol. 5, 137–141. https://doi.org/10.1016/j.cirpj.2012.03.003

Hu, K., Wu, A., Wang, H., Wang, S., 2016. A new heterotrophic strain for bioleaching of low grade complex copper ore. Minerals. 6, 12. doi:10.3390/min6010012.

Ibragimova, R.I., Mil'chenko, A.I., Vorob'ev-Desyatovskii, N.V., 2007. Criteria for choice of a brand of activated carbon for hydrometallurgical recovery of gold from ore pulps in carbon-in-leaching and carbon-in-pulp processes. Russ. J. Appl. Chem. 80, 891–903. https://doi.org/10.1134/S1070427207060092

Karigar, C.S., Rao, S.S., 2011. Role of microbial enzymes in the bioremediation of pollutants: A review [WWW Document]. Enzyme Res. https://doi.org/10.4061/2011/805187

Kim, Y., Seo, H., Roh, Y., 2018. Metal recovery from the mobile phone waste by chemical and biological treatments. Minerals 8, 8. https://doi.org/10.3390/min8010008

Krebs, W., Brombacher, C., Bosshard, P.P., Bachofen, R., Brandl, H., 1997. Microbial recovery of metals from solids. FEMS Microbiol. Rev. 20, 605–617. https://doi.org/10.1111/j.1574-6976.1997.tb00341.x

Kumar, R.N., Nagendran, R., 2007. Influence of initial pH on bioleaching of heavy metals from contaminated soil employing indigenous *Acidithiobacillus thiooxidans*. Chemosphere 66, 1775–1781. https://doi.org/10.1016/j.chemosphere.2006.07.091

Kushwah, A., Srivastav, J.K., Palsania, J., 2015. Biosorption of heavy metal: A review. Eur. J. Biotechnol. Biosci. 3, 51–55.

Lee, J., Pandey, B.D., 2012. Bio-processing of solid wastes and secondary resources for metal extraction – A review. Waste Manag. 32, 3–18. https://doi.org/10.1016/j.wasman.2011.08.010

Mehta, K.D., Pandey, B.D., Premchand, 1999. Bio-assisted leaching of copper, nickel and cobalt from copper converter slag. Mater. Trans. JIM 40, 214–221. https://doi.org/10.2320/matertrans1989.40.214

Mishra, D., Kim, D.J., Ralph, D.E., Ahn, J.G., Rhee, Y.H., 2007. Bioleaching of vanadium rich spent refinery catalysts using sulfur oxidizing lithotrophs. Hydrometallurgy 88, 202–209.

Mishra, D., Kim, D.J., Ralph, D.E., Ahn, J.G., Rhee, Y.H., 2008. Bioleaching of spent hydro-processing catalyst using acidophilic bacteria and its kinetics aspect. J. Hazard. Mater. 152, 1082–1091. https://doi.org/10.1016/j.jhazmat.2007.07.083

Mishra, D., Rhee, Y.H. 2010. Current research trends of microbiological leaching for metal recovery from industrial wastes, pp. 1289–1296. In Vilas, A.M. (ed.), Current Research, Technology and Education Topics in Applied Microbiology and Microbial Biotechnology, Volume 2, Formatex Research Center, Spain.

Nancharaiah, Y.V., Mohan, S.V., Lens, P.N.L., 2016. Biological and bioelectrochemical recovery of critical and scarce metals. Trends Biotechnol. 34, 137–155. https://doi.org/10.1016/j.tibtech.2015.11.003

Nedwell, D.B., 1999. Effect of low temperature on microbial growth: lowered affinity for substrates limits growth at low temperature.

FEMS Microbiol. Ecol. 30, 101–111. https://doi.org/10.1111/j.1574-6941.1999.tb00639.x

Nemati, M., Lowenadler, J., Harrison, S.T., 2000. Particle size effects in bioleaching of pyrite by acidophilic thermophile *Sulfolobus metallicus* (BC). Appl. Microbiol. Biotechnol. 53, 173–179. https://doi.org/10.1007/s002530050005

Niu, Z., Huang, Q., Wang, J., Yang, Y., Xin, B., Chen, S., 2015. Metallic ions catalysis for improving bioleaching yield of Zn and Mn from spent Zn-Mn batteries at high pulp density of 10. J. Hazard. Mater. 298, 170–177. https://doi.org/10.1016/j.jhazmat.2015.05.038

Niu, Z., Zou, Y., Xin, B., Chen, S., Liu, C., Li, Y., 2014. Process controls for improving bioleaching performance of both Li and Co from spent lithium ion batteries at high pulp density and its thermodynamics and kinetics exploration. Chemosphere 109, 92–98. https://doi.org/10.1016/j.chemosphere.2014.02.059

Ojuederie, O.B., Babalola, O.O., 2017. Microbial and plant-assisted bioremediation of heavy metal polluted environments: A review. Inter. J. Environ. Res. Public. Health. 14(12), 1504. https://doi.org/10.3390/ijerph14121504

Pathak, A., Healy, M.G., Morrison, L., 2018. Changes in the fractionation profile of Al, Ni, and Mo during bioleaching of spent hydroprocessing catalysts with *Acidithiobacillus ferrooxidans*. J. Environ. Sci. Health Part A Tox. Hazard. Subst. Environ. Eng. 53, 1006–1014. https://doi.org/10.1080/10934529.2018.1471033

Pathak, A., Morrison, L., Healy, M.G., 2017. Catalytic potential of selected metal ions for bioleaching, and potential techno-economic and environmental issues: A critical review. Bioresour. Technol. 229, 211–221. https://doi.org/10.1016/j.biortech.2017.01.001

Pradhan, D., Mishra, D., Kim, D.J., Ahn, J.G., Chaudhury, G.R., Lee, S.W., 2010. Bioleaching kinetics and multivariate analysis of spent petroleum catalyst dissolution using two acidophiles. J. Hazard. Mater. 175, 267–273. https://doi.org/10.1016/j.jhazmat.2009.09.159

Pradhan, D., Mishra, D., Kim, D.J., Chaudhury, G.R., Lee, S.W., 2009. Dissolution kinetics of spent petroleum catalyst using two different acidophiles. Hydrometallurgy 99, 157–162. https://doi.org/10.1016/j.hydromet.2009.07.014

Qu, Y., Lian, B., 2013. Bioleaching of rare earth and radioactive elements from red mud using *Penicillium tricolor* RM-10. Bioresour. Technol. 136, 16–23. https://doi.org/10.1016/j.biortech.2013.03.070

Rasoulnia, P., Mousavi, S.M., 2016. Maximization of organic acids production by *Aspergillus niger* in a bubble column bioreactor for V and Ni recovery enhancement from power plant residual ash in spent-medium bioleaching experiments. Bioresour. Technol. 216, 729–736. https://doi.org/10.1016/j.biortech.2016.05.114

Rivas-Castillo, A.M., Gómez-Ramirez, M., Rodríguez-Pozos, I., Rojas-Avelizapa, N.G., 2018. Bioleaching of metals contained in spent catalysts by *Acidithiobacillus thiooxidans* DSM 26636. Int. J. Biotechnol. Bioeng. 12(11), 430–434. https://doi.org/10.5281/zenodo.2021685

Rojas-Avelizapa, N.G., Hipólito-Juárez, I.V., Gómez-Ramírez, M., 2018. Biological treatment of coal combustion wastes by *Acidithiobacillus thiooxidans* DSM 26636. Mex. J. Biotechnol. 3, 54–67. https://doi.org/10.29267/mxjb.2018.3.3.54

Rojas-Avelizapa, N.G., Gómez-Ramírez, M., Alamilla-Martínez, D.G., 2015. Metal removal from spent catalyst using *Microbacterium liquefaciens* in solid culture. Adv. Mater. Res. 1130, 564–567. https://doi.org/10.4028/www.scientific.net/AMR.1130.564

Rousk, J., Brookes, P.C., Bååth, E., 2009. Contrasting soil pH effects on fungal and bacterial growth suggest functional redundancy in carbon mineralization. Appl. Environ. Microbiol. 75, 1589–1596. https://doi.org/10.1128/AEM.02775-08

Sag, Y., Kutsal, T., 2001. Recent trends in the biosorption of heavy metals: A review. Biotechnol. Bioprocess Eng. 6, 376–385. https://doi.org/10.1007/BF02932318

Santhiya, D., Ting, Y.-P., 2005. Bioleaching of spent refinery processing catalyst using *Aspergillus niger* with high-yield oxalic acid. J. Biotechnol. 116, 171–184. https://doi.org/10.1016/j.jbiotec.2004.10.011

Shailesh, R.D., Monal, B.S., Devayani, R.T., 2016. E-Waste: Metal pollution threat or metal resource? J. Adv. Res. Biotechnol. 1, 1–14. https://doi.org/10.15226/2475-4714/1/2/00103

Sood, G., Chitre, N., 2017. Review of biological methods for hazardous waste treatment. IOSR J. Biotechnol. Biochem. 06, 50–55. https://doi.org/10.9790/264X-06025055

Sukla, L.B., Esther, J., Panda, S., Pradhan, N., 2014. Biomineral processing: A valid eco-friendly alternative for metal extraction. Res. Rev. Microbiol. Biotechnol. 3(4), 9.

Suzuki, I., Lee, D., Mackay, B., Harahuc, L., Oh, J.K., 1999. Effect of various ions, pH, and osmotic pressure on oxidation of elemental sulfur by *Thiobacillus thiooxidans*. Appl. Environ. Microbiol. 65, 5163–5168.

Tay, J.H., Tay, S.T.L., Ivanov, V., Hung, Y.T., 2004. Apllication of biotechnology for industrial waste treatment. Handbook of Industrial Wastes Treatment. 2nd edition. Marcel Dekker, New York.

Vijayaraghavan, K., Yun, Y.-S., 2008. Bacterial biosorbents and biosorption. Biotechnol. Adv. 26, 266–291. https://doi.org/10.1016/j.biotechadv.2008.02.002

Wang, J., Bai, J., Xu, J., Liang, B., 2009. Bioleaching of metals from printed wire boards by *Acidithiobacillus ferrooxidans* and *Acidithiobacillus thiooxidans* and their mixture. J. Hazard. Mater. 172, 1100–1105. https://doi.org/10.1016/j.jhazmat.2009.07.102

Wang, J., Chen, C., 2006. Biosorption of heavy metals by *Saccharomyces cerevisiae*: A review. Biotechnol. Adv. 24, 427–451. https://doi.org/10.1016/j.biotechadv.2006.03.001

Xin, B., Jiang, W., Li, X., Zhang, K., Liu, C., Wang, R., Wang, Y., 2012. Analysis of reasons for decline of bioleaching efficiency of spent Zn-Mn batteries at high pulp densities and exploration measure for improving performance. Bioresour. Technol. 112, 186–192. https://doi.org/10.1016/j.biortech.2012.02.133

Xu, T.-J., Ting, Y.-P., 2009. Fungal bioleaching of incineration fly ash: Metal extraction and modeling growth kinetics. Enzyme Microb. Technol. 44, 323–328. https://doi.org/10.1016/j.enzmictec.2009.01.006

Yao, Z.T., Xia, M.S., Sarker, P.K., Chen, T., 2014. A review of the alumina recovery from coal fly ash, with a focus in China. Fuel. 120, 74–85. https://doi.org/10.1016/j.fuel.2013.12.003

Zhang, H., Yao, Q., Zhu, Y., Fan, S., He, P., 2013. Review of source identification methodologies for heavy metals in solid waste. Chin. Sci. Bull. 58, 162–168. https://doi.org/10.1007/s11434-012-5531-2

Zhuang, W.-Q., Fitts, J.P., Ajo-Franklin, C.M., Maes, S., Alvarez-Cohen, L., Hennebel, T., 2015. Recovery of critical metals using biometallurgy. Curr. Opin. Biotechnol. 33, 327–335. https://doi.org/10.1016/j.copbio.2015.03.019

4

Industrial Biotechnology and its Role in the Mining Industry

**Cuauhtémoc Contreras Mora[1], Susana Citlaly Gaucin Gutiérrez[1,*],
Hiram Medrano Roldán[1], Damian Reyes Jáquez[1],
David Enrique Zazueta Álvarez[1], Grisel Fierros Romero[2]
and Luis J. Galán Wong[3]**

[1]Instituto Tecnológico Nacional de México-Instituto Tecnológico de Durango, Unidad de Posgrado, Investigación y Desarrollo Tecnológico (UPIDET), Avenida Felipe Pescador 1830 Ote. 34080 Durango, Durango, México
[2]Tecnológico de Monterrey, School of Engineering and Sciences, Campus Querétaro, Avenida Epigmenio González No. 500, San Pablo, 76130, Querétaro, México
[3]Instituto de Biotecnología, Universidad Autónoma de Nuevo León, Avenida Manuel L. Barragán, S/N Cd. Universitaria, San Nicolás de los Garza, N. L, México
E-mail: susana_gaucin@hotmail.com
*Corresponding Author

4.1 Introduction

Microbiological leaching of low-grade ores for the recovery of various metals in solution has been practiced for quite some time. More recently, a variety of research projects has been directed toward the study of leaching of high-grade concentrates under controlled conditions. The rationale for such studies is that this kind of leaching could be competitive with smelting. Microbiological leaching lends itself to smaller plants, would be less energy intensive, and would result in less air pollution problems.

The ability of some microorganisms to assist in the recovery or remotion of metals from ores or tailings is now well known and their contribution to the solubilization of metal sulfides, while the mechanism is not completely clear,

73

is accepted and well documented. They are the only microbes to be used for metal recovery or removal on a commercial scale.

In general, microbes are ubiquitous on this planet and are capable of producing many varied metabolites under different environmental conditions. It has been shown that the mobilization of metals can result from microbial activity. The extent and magnitude of these "natural" phenomena are not well documented for at least two reasons. Many Limiting factors lead to very slow reaction kinetics and the amounts of metals involved are very small and the time span is long when considered against a man's working life. However, the amounts and effects can be large when measured in geological time. What are the reactions taking place in the natural environment? What are the limiting factors? Can these factors be optimized so that the ability of microbes to mobilize metals can be harnessed for the benefit of man? These are far-reaching questions and cannot be yet answered (Acevedo and Gentina, 2005).

Research on industrial biotechnology pretends to show, especially to the students' community, the virtues of using biotechnology in the productive mining sector, mainly due to its environmental problems, which implicate the presence of arsenic, manganese, bismuth, zinc, in concentrates as well as mineral tailings. Also, it is desired to discuss the importance of how to "sell" alternatives to the mining issue through research projects that allow research institutes to show, with evidence, their extension capacity with productive sectors. El arsenic can be found as a pollutant at levels up to 10%, appearing in valuable metal concentrates such as gold, silver, zinc, and lead among others, representing a conflict whenever such concentrates move a melting stage, causing detriments inside the ovens' walls, bedsides generating vapors containing arsenic. Manganese, which is found as a pollutant at levels up to 1%–1.5% in concentrates and mining tailings, forces to use more sodium cyanide in precious metals' recovery. The residues end up at the tailings' damps as residual cyanide, reaching elimination levels of this material above 90% at lab scale as well as pilot plant scale, including experiences at mine level in a rotatory reactor for manganese. All the researches using native *Thiobacillus ferrooxidans* strains have generated human resources at masters and doctoral levels in the bioengineering area, who, mostly, are currently working in the mining industry as well as generating patents of industrial interest (Medrano-Roldan et al., 2013a).

In these preliminary studies carried out at Durango Institute of Technology in the 1990s, some mining companies, and the Nuevo Leon

University, reported in several publications, further evidence was sought to reinforce these ideas. The technical feasibility for metal solubilization by microbes other than *Thiobacilli* was studied at laboratory scale. It is emphasized that conditions were not optimized, and economic viability was not sought. At the present time, the authors continue being well aware of the many problems in bridging the gap between technical feasibility and economic/commercial viability.

Growing attention is being given worldwide to the exploitation of deposits of refractory sulfidic silver and gold ores mainly, and bacterial leaching, under a biotechnology approach, is receiving increasing consideration as a technically and economically acceptable alternative to the conventional roasting process in the beneficiation of these ores (Haines, 1996; Medrano-Roldan et al., 2013a).

A number of long-term laboratory and pilot plant scale feasibility studies, including economic evaluations of bacterial leaching, have been reported in the literature (Medrano-Roldan et al., 1992). However, it is likely that the acceptability of bacterial leaching in the mining industry would be enhanced if there was a clearer definition of the process parameters, particularly those associated with particle size, mixing conditions, mineral concentration, and reaction rate, a knowledge of which is a prerequisite for rational scale up of the process (Lakshmanan, 1996).

In fact, heap and dump bacterial leaching of low-grade ores and residues are characterized by the low degree of control that can be exerted on the different process variables such as temperature, pH, aeration, mixing, nutrient concentration, microbial population, and others. This situation results in suboptimal rates of leaching and very prolonged processing periods (Acevedo et al., 1998).

The use of reactors offers the possibility of achieving a close control of the process, with the subsequent improvement of the overall productivity. However, it should be mentioned that direct evidence for poor bulk mixing in bioreactors comes mostly from ferric iron measurements during the propagation of microorganisms on minerals in order to recover valuable metals. Some authors (Acevedo et al., 1998; Deveci et al., 2003) have reported poorly circulating regions in which dissolved oxygen concentrations approached zero while high concentrations were maintained in the impeller zones in such a way that it would be interesting to study the use of air-sparged concentrated mineral pulp suspensions to correlate terminal blending time to an ion ferric production rate in terms of impeller geometry and silver and gold recovery.

In the case of copper biological leaching from chalcopyrite concentrates, the process has been improved through the use of several chemical compounds and environmental and operating conditions at shake flask and tank leaching where technical and economic parameters should be determined (Medrano-Roldan et al., 1992; Salazar, 1999). On the other hand, it is well established that the microbial technique for extraction of copper from chalcopyrite concentrates is associated with the transfer of oxygen, which is the oxidizing agent, and carbon dioxide which is the source of carbon. The purpose of aeration and agitation during leaching is to achieve these gas transfers and to mix the solid substrate and nutrient solution in such a way that a uniform suspension of chalcopyrite particles and bacteria can be maintained. There are several methods which have been successfully applied to improve the aeration and agitation of the leach solution, for example, air spargers, magnetic stirrers, and reciprocating shakers. Some authors have shown that the best results for copper extraction were obtained with techniques having the highest rate of oxygen and carbon dioxide mass transfer into the leach solutions. In the agitated leach solutions, the surface of the solution is continuously renewed and the oxygen transfer is therefore accelerated. In this category are the shake flask techniques. In addition, copper leaching operations can be carried out by using *Thiobacillus ferrooxidans* at an ambient temperature in substrates such as mine wastes and concentrates with the addition of ferrous iron in order to enhance the bioleaching process.

Biohydrometallurgical extraction of copper from chalcopyrite can be described by the following electromechanical reactions (Salazar, 1999):

Equation 1. Reaction anodic

$$2CuFeS_2 + 16H_2O + H_2SO_4 \rightarrow 2Cu_2 + 2Fe^{3+} + 5SO_4^{2-} + 34H^+ + 34e^-$$

Equation 2. Reaction cathodic

$$34H^+ + 34e^- + 81/2\ O_2 \rightarrow 17H_2O$$

Equation 3. Reaction general

$$2CuFeS_2 + 81/2\ O_2 + H_2SO_4 \xrightarrow{\text{(Bacteria)}} 2CuSO_4 + Fe_2\ (SO_4) + H_2O$$

The above reactions are consistent with the semiconductor nature of copper sulfides and their electromechanical degradability (Silverman and Erlich, 1994). This electromechanical characteristic of copper sulphides was subject of leaching studies, of normal composition. There is much information is available on the bioleachabvility of chalcopyrite concentrate but the process not always is progressing with the same sentido, suggesting that chalcopyrite may be made of two different mineralogical compositions. The one which is easily leached by the microbes has a normal composition. This is built of a cupric, a ferrous, and two sulfide ions. The chalcopyrite, which is difficult to leach, is called abnormal. This may be composed of a cuprous, a ferric, and two sulfide ions. Besides, the mineralogical identification of these two different chalcopyrite species is not possible since they give the same spectra; so the difference in the purity (chemical composition) of different chalcopyrite may also influence their leachability by the microorganisms.

There are several mining companies interested in evaluate using baffled flasks in order to improve oxygen and carbon dioxide transfer conditions, temperatures and pH values for removing contaminants from mineral containing gold, silver, copper, and others, these contaminants could be arsenic, manganese, bismuth, zinc, etc. Besides, several impeller designs could be tested at well-mixed batch reactor level.

Looking at the gas–liquid mass transfer considerations is very important in the choice of experimental equipment for kinetic experiments including the scale-up process. As was mentioned, in the bacterial oxidation reaction of mineral sulfides, oxygen and carbon dioxide are consumed. Therefore, usually, biooxidation experiments on mass transfer limitation of oxygen and carbon dioxide from the gas-phase (which is usually air) to the liquid phase need to be prevented; the maximum oxygen transfer rate in the slurry needs to exceed the consumption rate, and similarly, for carbon dioxide ("Course and Application of Biotechnology," 1997).

The bacterial oxidation of sulfide minerals is a surface reaction and the reaction rate is proportional to the surface area of the mineral. Besides, the oxygen and carbon dioxide consumption rates are directly related to the bacterial oxidation rate of a metal sulfide. From the stoichiometry of the chemical oxidation of pyrite, it follows that 1 mol of pyrite requires 3.75 mol of O_2. In the bacterial oxidation reaction, this stoichiometry coefficient is slightly different in the autotrophic biomass production in such a way that this difference can be neglected. On the other hand, it should be considered that molecular oxygen acts as the final electron acceptor in the energy-yielding reactions involved in bacterial leaching, i.e., the oxidation of ferrous ion, reduced sulfur, or elemental sulfur.

Thus, if one desires to study a bioleaching process from the mass transfer viewpoint, it should be considered that the oxygen demand of a microbial culture can be related to the specific growth rate μ which can be taken as a function of the Fe^{3+} production, in such a way that it can be written as follows:

$$\textbf{Na} = \boldsymbol{\mu} \ \textbf{X/Yo},$$

X being the biomass concentration. Oxygen cell yield (**Yo**) can be determined experimentally or estimated through a mass balance.

Oxygen and CO_2 must be supplied to the bacteria at rates at least equal to its demand. If that is not the case, the cells will grow under the limitation of both compounds; so the growth will be linear instead of exponential and severe damage may occur at the electron transport and oxidative phosphorylation level.

Oxygen supply, or oxygen transfer rate (OTR), is given by:

$$\text{OTR} = \text{Kla } (\text{C*–Cl}).$$

The transfer potential, **C*–Cl**, is limited by the low solubility of oxygen in aqueous media, impeller design, aeration and agitation levels, temperature, and the rheology of the medium. In the same way is the case of the volumetric transfer coefficient **Kla**.

As we can see, the application of the biotechnology knowledge in the mining industry is complex, specifically when all the problems related to the chemical engineering field must be solved. Recently, many industries have shown interest in the use of *Thiobacillus ferrooxidans* as a possible biological solution to several bioleaching processes as an alternative to the traditional chemical methods, such as bioleaching of metals, treatment of acid mine drainage, elimination of sulfur from sour gases, and removal of hydrogen sulfide from industrial and municipal effluent gases. However, in most of these application areas, the results obtained have not been sufficiently satisfactory to bring about the necessary adaptation of the potential of this microorganism from industrial use (Valayapetre and Torma, 1988).

The first step towards an effective industrial of bioleaching is the study of the kinetics of oxidation maintained by the microorganisms in such a way bacterial leaching is already a commercially successful process used for pre-treatment of refractory gold-bearing sulfides. The first commercial operation, at Fairview Gold Mine (South Africa), began in 1986 with a capacity of 10 tons per day (Boogerd et al., 1990). This plant now treats 35 tons of a pyrite arsenopyrite concentrate per day. The largest bacterial leaching plant, at Ashanti in Ghana, presently treats 1100 tons of gold bearing pyritic concentrate per day. There are also operations at Sao Bento in Brazil and at Harbour Lights in Australia (Boogerd et al., 1990). On the other hand, an interesting industrial bacterial leaching project was set up in Uganda to extract cobalt from a cobaltiferous pyrite ore (Jensen and Webb, 1995).

At the end of the 1990s, in Mexico, several research groups at the National Polytechnic Institute, Mexico Autonomous National University, Nuevo Leon Autonomous University, and Durango Institute of Technology, among others, including some mining companies, have been working and looking for alternatives in order to solve problems involved in this industrial sector especially those related to removal of arsenic sulfide from gold–silver and lead–zinc concentrates and copper from chalcopyrite concentrate and, more recently, manganese from mineral containing gold and silver improving the use of sodium cyanide in the recovery system of these metals as soon as sodium cyanide regeneration from mines effluents (Dew et al., 1998; D'Hughes et al., 1997; "IMPI-SECOFI 950410," n.d.).

Manganese occurs in nature in both the oxidized and the reduced form, as well as several low-grade oresits solubilization can be carried out through microorganisms isolated from mines drainages in such a way an industrial process can be designed in order to reduce the use of sodium cyanide during the gold and silver recovery process.

One way to recover manganese from low-grade ore is its solubilization through sulfation roasting and subsequent water leaching, but the proposed processes do not seem to have found any commercial application. At neutral pH, the most likely condition occurring in dumps, manganese oxides with manganese of oxidation states ($^{3+}$) and ($^{4+}$) are insoluble, whereas reduced manganese might be the microbially mediated reduction of its oxides.

Finally, and generally speaking, it is apparent that tailing dams and waste deposits represent extremely complex and variable environments. All mineral deposits differ in composition and extensive mineralogical and grade variations can exist within any one deposit. In mining practice, attempts are often made to blend feed ores in an effort to maintain uniform grades. Despite this, variations in the nature of the tailings occur between different sections of the same dam as a result of changes in the grade of the run-of-mine feed, fluctuations in metallurgical extraction efficiency, and physical characteristics of the dam including water availability and depth. The residence time of the deposited material must also be considered, and such a complexity does not permit uniform modeling of sulfide mineral dissolution and bacteria oxidation of mine tailings. Nevertheless, significant progress has been made in our understanding of the chemical and biological processes associated with these materials. This must contribute to the attempts to industrially exploit the sizeable metal values contained within these low-grade deposits and to minimize the possibilities of pollution.

4.2 Application of Bioleaching in Industrial Problems with Zn and Bi

The new industrial process of mining demands the resolution of problems focused in the environment (Brüger et al., 2018; Lin et al., 2019); in this case, the bioleaching is the available alternative for the resolution of mineral problems to different points of the process and in the improvement of final products of metal to obtain a better price in the market with the cost–benefit. Some improvements in the metals are the elimination of metals of low prices that act like a pollute; such is the case of the metal of bismuth that is a pollute of a product of copper, gold, and silver. This product of a mining corporation in the north of the city of Durango is called "Minera de Avino". The bismuth is present in a concentration of 2% of the final product; this metal decreases the price. The bismuth is a metal hard to remove with conventional methods. The industry needs to remove at least the 50% total of bismuth, this problem can be resolved with help of bioleaching, this study was performed with native bacteria thus was obtained of ground of the mining process, these bacteria were propagated in baffled flasks of 500 mL, with 20 g of minerals and 160 mL of 9 K medium (modified) for four days until the cellular density was of 1×10^7 cell/mL.

The experiments were performed with native bacteria in baffled flasks of 500 mL, with 20 g of minerals and 160 mL of 9 K medium (modified) with four different pH (Table 4.1); each experiment was performed with three repetitions and environment of 30°C and agitation of 160 rpm.

The bacteria in the flasks was monitored with Neubauer camera in a Leica CME microscopy in the beginning and at the end of each experiment. The parameters of pH and oxidation–reduction (Redox) was measured used a multiparameter Hanna 991003 at the beginning and in the end of each experiment (Table 4.2).

The rise in the pH was the consequence for the demand of protons of hydrogen, essentials in the bacteria metabolism to obtaining energy from acid in the medium; this decrease has influence in the solubilization of bismuth

Table 4.1 Experiment of remotion of bismuth with different levels of pH

Experiment	pH Level
Guerrero 1	1.5
Guerrero 2	2
Guerrero 3	2.5
Guerrero 4	3

Table 4.2 Results of bioleaching

Experiment	pH Initial	pH Final	Redox Initial	Redox Final	Bacteria Initial (Cell/mL)	Bacteria Final (Cell/mL)	Final Concentration of Bismuth (%)
Guerrero 1	1.5	3.1	406	316	1.1×10^7	1.6×10^7	1.55
Guerrero 2	2	3.1	328	285	1.1×10^7	1.9×10^7	1.30
Guerrero 3	2.5	4.2	354	329	1.1×10^7	1.7×10^7	1.40
Guerrero 4	3	4.3	343	321	1.1×10^7	1.5×10^7	1.86

because the acid pH played an important role in the retention of metallic ions in the medium (Wang et al., 2018); in this case, the solubilization provoked a decrease in the potentials of oxidation–reduction because the medium was rich in ions of metals (Schippers et al., 2019); because the bacteria used the bismuth like a donator of ions in their breathing chains, the authors have indicated that the relation of this separation occurred in the biofilms, and they also give support and resistance to different metals in hostile environments (Liu et al., 2019).

The biological process with pH 2 obtained the best remotion of bismuth with 0.7% of bismuth, but the company needs only the 1% of total, becoming in 70% of remotion; this process did not take other metals like copper, gold, and silver, increasing the price of the final product (Jeon et al., 2017).

The bioleaching has found a different solution of industrial problems in the products of minerals wasted in the case of elimination of heavy metals for an extra benefit (Wang et al., 2015; Kim et al., 2016). This kind of experiments was done for the elimination of zinc that pollutes mineral wastes with a lot of metals of iron and decreases the value of this metal. The elimination with bioleaching was done in baffled flasks of 500 mL, with 20 g of minerals, and 160 mL of 9 K medium (modified) and was adjusted to pH 2 every 24 h and incubated in a temperature of 30°C and a shaker of 160 rpm. The bacteria in the flasks were monitored with Neubauer camera in a Leica CME microscopy. The parameters of pH and oxidation–reduction (Redox) and a multiparameter Hanna 991003 were used once per day in each experiment. Each experiment was performed in triplicate for 19 days.

Since the first day, the bacteria take the different metallic components for its metabolism, and in the seven days, the bacteria population had an adaptation to the medium and the metals; this adaptation was fast in comparison with commercial bacteria until they achieved the maximum growth in day 10 (Blais et al., 1993; Elzeky and Attia, 1995).

The oxidation was raised because the bacteria used the sulfate ferrous in the production of energy following the equation of Suzuki (2001):

Equation 4. Oxidation use sulphate ferrous

$$FeS_2 + 6Fe^{3+} + 3H_2O \rightarrow S_2O_3^{2-} + 7Fe^{2+} + 6H^+$$

$$S_2O_3^{2-} + 8Fe^{3+} + 5H_2O \rightarrow 2SO_4^{2-} + 8Fe^{2+} + 10H^+$$

These reactions raised the potentials of oxide-reduction since the first day of experimentation (Figure 4.1), and with the generation of sulfuric acid in the metabolism of the bacteria, the zinc was converted in a form more soluble in the medium (Natarajan, 2018) following the reaction (5).

Equation 5. Bioleaching of Zn

$$H_2SO_4 + Zn \rightarrow H_{2(gas)} + ZnSO_4$$

The adjustment of acid helped in the extra disposition of acid and played as an essential parameter for the solubilization of zinc (Ma et al., 2017) and the materials of carbonates and silicates present in the metal which helped in the solubilization of zinc (Shen et al., 2018).

The experiment shows that zinc was removed at 60% which gives value to this waste for to the mining industry with an environment friendly process.

4.2.1 Biooxidation of Ions of Ferrous Fe^{2+}

In strong reduction conditions, the proton concentration controls the pyrite leaching according to the reactions proposed by Juárez (2004).

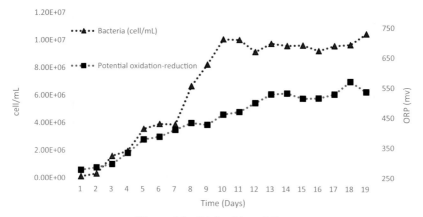

Figure 4.1   Bioleaching of Zn.

Equation 6. Pyrite general reduction

$$FeS_2 + 2H^+ + 2e^- \rightarrow Fe^{2+} + 2HS^-$$

The standard redox potential in the aerobic bioleaching of Mn is greater than that of Fe since the valences of these in the anaerobic bioleaching are Mn^{+4}/Mn^{+2} and Fe^{+3}/Fe^{+2}; this is given by the following reactions (Zachariades and Fraser, 1991; Cornell and Schwertmann, 2003).

Equation 7. Biooxidation of pyrite and formation of ferric sulfate and sulfuric acid

$$2FeS_2 + 2H_2O + 7O_2 \underrightarrow{Bacterias} 2FeSO_4 + 2H_2SO_4$$

Equation 8. General reduction of manganese (IV) to manganese (II)

$$Mn^{+4} + Fe^{+2} \rightarrow Mn^{+2} + Fe^{+3}$$

In the anaerobic bioleaching of the pyrolusite, microorganisms are able to use the pyrolusite as the final electron receptor in the chain of reactions to the metabolism of the bacteria *A. ferrooxidans* with the pyrite (Zhang et al., 2008; Syed, 2012).

Equation 9. Oxidation of pyrite to Fe^{2+}

$$FeS_2 + 6Fe^{+3} + 3H_2O \rightarrow S_2O_3^{-2} + 7Fe^{+2} + 6H^+$$

Equation 10. Reduction of Fe^{3+}

$$S_2O_3^{-2} + 8Fe^{+3} + 5H_2O \rightarrow 2SO_4^{-2} + 8Fe^{+2} + 10H^+$$

Equation 11. Oxidation of Fe^{2+}

$$8Fe^{+2} \rightarrow 8Fe^{+3} + e^-$$

Equation 12. Oxidation of pyrolusite with electron flow re-oxidation reactions and reduction of Fe

$$MnO_2 + 4H^+ + 2e^- \rightarrow Mn^{+2} + 2H_2O$$

Describing thus the decomposition of the pyrolusite and releasing the minerals within the matrix to be able to recover them according to the reactions proposed by Li (2005).

4.3 Bioconversion Evaluation of Ferrous Ions

The isolation of microorganisms from mine material is possible with the use of selective media, as is the case with modified 9 K culture medium (Medrano-Roldan et al., 2013b) and culture medium MR (Greene et al., 1997), where the use of microbial consortiums obtained from mine material can have a broad industrial use (Table 4.3). In this case, the use and mobility of iron in the MR culture medium is evaluated, which promotes reducing processes and the 9 K that promotes oxidation processes.

Each of the samples was stimulated for microbiological growth using 15% (w/v) of sample in 250 mL Erlenmeyer flask with MR and 9 K, respectively (Table 4.4). They were incubated at 33°C, 150 rpm and pH measurements were made, redox potential with multiparameter Hanna 99103 and cellular counting in Leica microscope.

The consortiums are classified according to their morphological characteristics in colony; in Table 4.5, the bioleaching processes of iron are analyzed as hematite in a concentration of 18% (w/w) present in the tailings, where the reaction carried out by the microorganisms in the 9K culture medium is from reduction of ferric ions (Fe^{3+}) to ferrous ions (Fe^{2+}) reaction described by Equation (13) to subsequently form ferric ions, where it is mineralized insolubly in the form of goethite according to the reaction of Equation (13) postulated by Dean (1999).

Table 4.3 Concentration of iron in environmental samples of mine

Sample of Mine Material	Fe (ppm)
Mine blue lighthouse	30,039.6
New powder	71,143.9
Old powder	84,042.2
Vegetal soil	67,683.3

Table 4.4 Bacterial consortiums obtained

Strain	Sample Origin	Medium
3	New powder	9 K
Red	New powder	9 K
C.P	Water	9 K
C.P.R.	Blue lighthouse	9 K
1	Blue lighthouse	MR
B-2	Mine tailings	MR
Reddish	Mine tailings	MR
R-2	New powder	MR

Table 4.5 Bioleaching of Fe

Microbial Consortium	Fe Day 1 mg/L	Fe Day 10 mg/L	ppm Bioleached of Fe	Bioleached (%)
Analytical white MR	101,500	95,800	5700	5.6%
Analytical white 9 K	101,500	115,900	−14400	−14.2%
3 (9 K)	101,900	117,500	−15,600	−15.3%
C.P.R (9 K)	113,200	118,700	−5500	−4.9%
Red (9 K)	122,700	120,800	1900	1.5%
1 (MR)	109,000	108,000	1000	0.9%
B2 (MR)	103,200	106,800	−3600	−3.5%
Red 2C (9 K)	103,500	109,800	−6300	−6.1%
Reddich (MR)	108,500	111,400	−2900	−2.7%
R-2 (MR)	106,200	104,300	−1900	1.8%
3 (MR)	94,900	101,000	−6100	−6.4%

Equation 13. Oxidation of Fe^{2+}

$$Fe^{+2} + nH_2O \rightarrow Fe(OH)_n^{2-n} + nH^+$$

Table 4.5 shows that a greater flow of conversion of ferrous to ferrous ions for the formation of goethite is with inoculum 3 of the culture coming from medium 9 K where the iron fixation is 15,600 ppm, in comparison with the consortia of the MR culture medium, where the reduction of ferric ions results in more soluble ferric ions in the medium.

4.4 Use of Bioleaching in Refractory Materials of Mn and S

Among the techniques used, with viable economic characteristics for the treatment of these soils is heap leaching, where the material after the grinding process is transported (usually by conveyor belts) to the place where the pile will be formed. During this journey, the material is subjected to first irrigation with a solution of water and acid solutions, known as the curing process. In its destination, the ore is discharged by means of gigantic spreading equipment, which deposits it neatly forming a continuous embankment of 6–8 m high, forming the leach pile. A drip irrigation system and sprinklers that cover the entire exposed area are installed on this pile. Under the piles of material to be leached, a waterproof membrane is installed beforehand, on which there is a system of drains (pipes with slots) that allows collecting the solutions that infiltrate through the material contained in the oxidized minerals, forming a solution with the minerals of interest. The leaching is maintained for 45–60 days, after which it is assumed that the amount of the mineral to be solubilized has been almost completely exhausted so that the remaining material (jal) is transported for a second leaching process and can extract the rest of the mineral sought (Beckel, 2000).

The speed of the leaching process is initially high since the reagent directly attacks the species present on the surface of the particle; with the passage of time, the speed of the reaction decreases due to the fact that the reaction surface is increasingly removed from the surface. The surface of the particle and the reactants and products take more time to move inside the particle. The heap leaching, which, from its first application for the recovery of gold from low-grade minerals by cyanidation in the 1970s, has, in conjunction with solvent extraction and electro-winning, the advantages to be a key hydrometallurgical technology for the recovery of basic metals, mainly copper from secondary oxides and sulfides (Padilla et al., 2008; Sylwestrzak, 2010).

On the other hand, there is also the technique of slag heaps when talking about processes with low operating costs. The tailings are heaps of ore of large dimensions in which minerals are accumulated at a very low grade, around 0.1%–0.5%, where suitable soils are used in which the leaching solution, as in leaching in piles, passes through the material recovering through the bottom of the waste; thus, the leaching solution is usually treated by electrolytic reduction, recovering the metal of interest. In this way, the leaching solution (refined) is recirculated to attack more minerals (Bouffard and Dixon, 2001). These processes used at the industrial level are carried out through the generation of chemical reactions, by the leaching solutions of chemical origin that are used.

The lack of efficient techniques that can carry out the total extraction of the metals of interest, either alone or in conjunction with conventional techniques, means that one of the alternatives with industrial applications is bioleaching, which, today, is one of the main operations of biotechnology applied to the extractive metallurgy industry (Morin et al., 2006). Bioleaching can be carried out in different ways, *in situ*, leaching applied to very low-grade minerals, <0.1%. This technique involves irrigating the walls of the mine with a biological leaching solution. The mineral must have been previously fractured or it must be a porous mineral. This process is used in mines where conventional mineral reserves have already been depleted (Morin et al., 2006).

Like the conventional process, this biotechnological process can be carried out by means of tailings,where the heaps of ore are disposed in suitable soils in which the leaching solution previously treated as a culture medium for the microorganisms is irrigated by the ore for its subsequent collection at the bottom of the heap (Bouffard and Dixon, 2001), the mineral being treated for the recovery of minerals by conventional methods since the solid matrix was fractured or solubilized by the action of microorganisms. The biological treatment on the surface of the mineral has a great impact when using bioleaching through heap leaching due to the large amounts of material that can be processed; this technique has been used at industrial level for the recovery of metals of economic importance, in particular, copper, gold, silver, and uranium, and this makes the technique with a high degree of selectivity in the mining industry. This method is for minerals with low grades, trying to make the mineral particle more uniform (Bouffard and Dixon, 2001).

The use of this type of techniques at industrial level is due to laboratory work, which has led to the determination of various variables for the understanding of the behavior of the microorganisms used during the biological processing of mineral samples; however, the importance of the experiments that are carried out at the laboratory level goes beyond the industrial application, but also of the generation of information of these biological systems, which not only depend on the nature of the sample but also on the operating conditions given during the process and the ability of microorganisms to adapt to the extreme conditions of the process environments. As it has been reported, the bioleaching success will depend of the following variables or conditions: there are reports in which residence times can range from 4 min (Bouffard, 2008) to 90 days, even 364 days (Hernández et al., 2007). The flow velocity of the leaching solution can vary and is a function of the particle size of the material by the compaction that is generated during wetting with the solution. The temperature can be a differential parameter on the type of microorganism species with which one is working (Sand et al., 1992); however, when using this type of bioleaching, the control of this variable is complicated, unlike in a bioreactor where the control of the temperature of the system is possible, although it has been demonstrated that the survival of the microorganisms can occur even in a temperature variation of up to 10°C (García-Moyano et al., 2008). Aeration can be controlled over the leaching solution, but already during the interaction with the ore in the land is complicated, so these processes can be considered microaerophilic and thus use some of the materials as a final electron acceptor in its chain of respiration (Sugio et al., 1988) promoting the elimination of some problem minerals.

The source of energy that can be various minerals such as iron and sulfur are essential for the development of microorganisms in the bioleaching processes, which may be on the material being treated; in this way, the microorganisms develop strategies to be able to acquire them, as is the production of enzymes, thus carrying the ion exchange on the mineral matrix. High Fe^{3+}/Fe^{2+} values in mine effluents and leachates are generally indicators of biological activity, as is the natural mechanism of iron oxidation in leachates or acid drainages (Guerinot, 1994). The adjustment of suitable pH value is important for the growth of the bacteria in leaching and is decisive for the solubilization of the metals. PH values in a range between 2.0 and 2.5 are optimal for the bacterial oxidation of ferrous iron and sulfides (Bosecker, 2001).

During the last years, in Mexico, several research groups have been working in search of alternatives to solve problems that arise in this industrial sector, especially for the removal of sulfur dioxide from gold–silver and lead–zinc concentrates, concentrates from copper, from chalcopyrite and manganese, and concentrates containing gold and silver, trying to improve recovery in conjunction with the use of sodium cyanide in the recovery systems of these metals.

The manganese solubilization can be carried out through the microorganisms isolation from the drainages of the mines in the form of an industrial process during the recovery of gold and silver. Manganese is solubilized as manganese (II) sulfate from manganese (IV) oxide by the action of bioleaching bacteria (Galen and Martinus, 1925).

Equation 14. Bioleaching of pyrolusite general and semi reactions

$$2MnO_2 + 4SO_2 + H_2O \rightarrow Mn_2(SO_3)_3 + H_2SO_4$$

$$Mn_2(SO_3)_3 \rightarrow MnS_2O_6 + MnSO_3$$

$$MnO_2 + H_2SO_3 \rightarrow MnSO_4 + H_2O$$

At a neutral pH, the manganese oxide and manganese are insoluble, and the reduced manganese is relatively mobile. Therefore, a possible form of manganese mobilization could be the microbial reduction of its oxides (Budinova et al., 2009). On the other hand, the recovery of manganese from these minerals has been determined by solubilizing it through the leaching of sulfated water, but the proposals do not seem to have found a commercial application. In this way, the use of a native consortium for the removal of manganese has been determined as a process with a view to commercial application. This process is of great importance since the manganese in the form of pyrolusite (MnO_2) prevents the recovery of silver, leaving considerable concentrations in the residues of the concentrates that were treated by cyanidation. The number of tons of this waste (mining tailings) makes the biological alternative by relatively simple methods attractive from the economic point of view.

In the Graduate Unit, Research and Technological Development (UPI-DET), belonging to the Technological Institute of Durango, the academic body of Innovation Research Projects in the Food and Biotechnology Area (IAB) has been in charge of supporting local agencies in the sector mining for the resolution of manganese problems. The methodology at the laboratory level to carry out the removal of manganese up to 67% and the subsequent recovery of silver up to 40% has been determined; it should be noted that this percentage of silver recovered was already considered lost.

In Figure 4.2, it is observed that the pH and the concentration of ferrous sulfate have an effect on the removal of manganese, reaching up to 67%. Methodologies have been reported for the elimination of manganese, where ferrous sulfate (50 g/L) can be used, temperatures in the range of 70°C with the addition of sugars generating a precipitation of manganese in the form of carbonates (Kane and Cardwell, 1974; Acharya et al., 2004). So the use of a biological system shows that with the addition of ferrous sulfate as an energy source, suitable conditions for microorganisms to have initial pH ranges from 2.0, promotes anion exchange on the surface of the mineral, so as to carry out the removal of manganese after having been modified in kind by the effect of

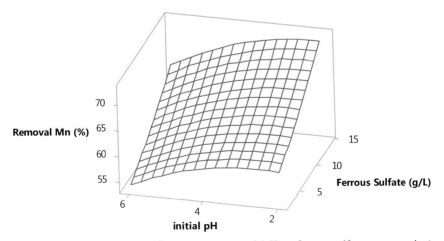

Figure 4.2   Surface response of manganese removal (pH vs. ferrous sulfate concentration).

bacterial metabolism as a sulfate (Das et al., 1998) at ambient temperatures. It is known that for the system to reach an initial pH of 2, the addition of sulfuric acid is considerable because one of the characteristic components of the pure jal is calcium carbonate, an agent with which the pH is adjusted to the cyanidation process, a potential expense is generated, but it can be seen that even adjusting the initial pH to 4 can obtain manganese elimination percentages greater than 60%, which is a positive factor for the economy of the process, avoiding excessive addition of sulfuric acid, and noting that the consortium can perform the required tasks in pH ranges higher than those reported in the literature.

In aerobic processes, the supply of gaseous substrates is important; however, during a heap leaching process, these substrates are deficient due to the nature of the system. In Figure 4.3, the concentration of ferrous sulfate exerts an effect on the elimination of manganese and it can be observed that a laboratory-level mixing does not influence, which determines that the process could be carried out in a microaerophilic environment at an industrial level where a mixing process would be difficult to perform; in this way, in an oxygen-poor environment, some bioleaching bacteria can use manganese (IV) as the last electron acceptor in their breathing chain (Haoran et al., 2005), transforming the pyrolusite matrix into a species that is mostly soluble due

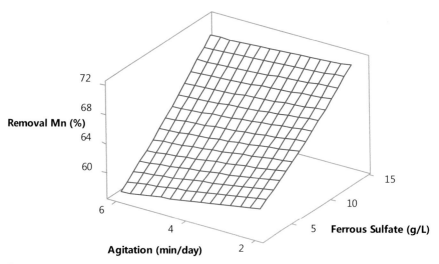

Figure 4.3 Surface response of manganese removal (agitation vs. ferrous sulfate concentration).

to the system conditions, allowing the subsequent action of sodium cyanide for the recovery of silver due to the fracturing of the solid matrix by the bioleaching processes.

These results show that the recovery of silver from waste may be possible due to the removal of manganese from the mineral surface or the possible fracturing of the solid matrix, which causes the efficient production of silver by the cyanidation process after biological treatment.

Figures 4.4 and 4.5 show the best percentages of silver recovery, denoting that the optimum pH range for the growth of microorganisms (1.5–2.0) negatively affects since the silver was leached from the solid phase; this may be due to the large amounts of sulfuric acid added in the initial pH adjustment, and the same acid generated by the metabolism of the bacteria. This effect was determined as undesired because there is no methodology for the precipitation of silver in the liquid phase by the action of acidic substances; however, it has been shown that the microbial action even when manganese has not solubilized can obtain positive silver extraction percentages.

In this way, the importance of laboratory-level tests is crucial for the development of methodologies with interest in the industrial application and the generation of information about academic research projects.

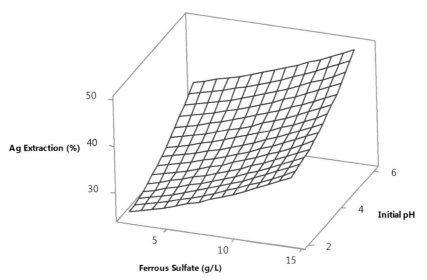

Figure 4.4 Surface response of silver extraction (pH vs. ferrous sulfate concentration).

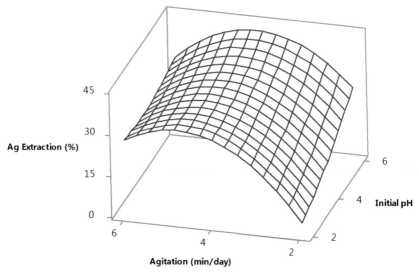

Figure 4.5   Surface response of silver extraction (pH vs. agitation).

4.5 Bioprocess Engineering and Economics

Finally, it is possible to establish that evaluation of the engineering economics of mineral bioprocesses is important as new products move from
the research and development phase into commercial production and test
marketing. Metals of current interest include a wide range of microorganisms
and its biochemical intermediates. Engineers and scientists sometimes need
to obtain a first estimate of relative costs to choose process alternatives, as
heap leaching for instance, and establish magnitude-of-order product costs.
Since there is such a broad range of metals, as recovery or remotion, with
different economic criteria, a single type of cost calculation approach cannot
be used for all the cases. Consequently, procedures to estimating the relative
economics for each type of metal on an internally consistent basis must be
used.

Mining bioprocess economics require the engineer or scientist to be
able to quickly recognize limiting technical parameters that are likely to
be encountered on scale-up. Processing schemes that minimize operational
difficulties and cost must be selected accordingly. Hence, process concep-
tualization is an important part of the evaluation process and must reflect
regulatory constraints as well as transport and kinetic phenomena required to
size key pieces of equipment. In the mining industry, examples are oxygen

and carbon dioxide transfer, mixing, and heat removal in a fermentation i.e., compressor and heat exchange costs: in some cases, it is difficult to carry out calculations since physical properties differ (e.g., broth viscosities and heat capacities), as a function of the pulp concentration, or diffusion coefficients in the particles of mineral for the heap leaching process where there exist several parameters which are not well defined during the early stages of a research and development program. Hence, approaches and heuristic rules that might be used in this situation are also needed. This chapter addresses the basics of industrial biotechnology applied to the mining industry.

References

Acevedo, F., Cassiuttolo, M.A., Gentina, J.C., 1988. Comparative performance of stirred and Pachuca tanks in the bioleaching of a copper concentrate. Biohydrometallurgy. STI, 385–394.

Acevedo-Bonzi, F., Gentina, J.C., 2005. Fundamentos y Perspectivas de las Tecnologías Biomineras. Archivos de Ingeniería Bioquímica. Pontificia Universidad Católica de Valparaiso, Chile.

Acharya, C., Kar, R.N., Sukla, L.B., Misra, V.N., 2004. Fungal leaching of manganese ore. Trans. Indian Inst. Met. 57, 501–508.

Beckel, J., 2000. "El proceso hidrometalúrgico de lixiviación en pilas y el desarrollo de la minería cuprífera en Chile", Santiago de Chile, NU. CEPAL. División de Desarrollo Productivo y Empresarial.

Blais, J.F., Tyagi, R.D., Auclair, J.C., 1993. Bioleaching of metals from sewage sludge: Microorganisms and growth kinetics. Water Research, 27(1), 101–110. https://doi.org/10.1016/0043-1354(93)90200-2

Boogerd, F.C., Boss, J.J., Heijnen, J.G., Lans, R.V.D., 1990. Oxygen and carbon dioxide mass transfer and the aerobic, autothrofic cultivation of moderate and extreme thermofiles: A case study related to the microbial desulfurization of coal. Biotechnol. and Bioeng. 35, 1111–1125.

Bosecker, K., 2001. Microbial leaching in environmental clean-up programmes. Hydrometallurgy, Elsevier, 59(2–3), 245–248.

Bouffard, S., 2008. Agglomeration for heap leaching: Equipment design, agglomerate quality control, and impact on the Heap leach process. Minerals Engineering. 21, 1115–1125.

Bouffard, S., Dixon, D., 2001. Investigative study into the hydrodynamics of heap leaching processes. Metallurgical and Materials Transactions B, Springer, 32(5), 763–776.

Brüger, A., Fafilek, G., Restrepo, O.J., Rojas-Mendoza, L., 2018. On the volatilisation and decomposition of cyanide contaminations from gold mining. Sci. Total Environ. 627, 1167–1173. https://doi.org/10.1016/J.SCITOTENV.2018.01.320

Budinova, T., Savova, D., Tsyntsarski, B., Ania, C., Cabal, B., Parra, J., Petrov, N., 2009. Biomass waste-derived activated carbon for the removal of arsenic and manganese ions from aqueous solutions. Elsevier. 255, 4650–4657.

Cornell, R.M., Schwertmann, U., 2003. "Adsorption of Ions and Molecules. The Iron Oxides. Structure, Properties, Reactions, Occurences and Uses". Wiley-VCH. 2, 253–296.

Course on Application of Biotechnology to economic recovery of metals from ores and concentrates, 1997. Australian Mineral Foundation. Sydney, Australia.

D'Hughes, P., Cezac., P., Cabral., T., Battaglia., F., Truong-Meyer, X.M., Morin, D., 1997. Bioleaching of a cobaltiferous pyrite a continuous laboratory scale study at high solid concentration. Miner. Eng. 10, 507–527.

Das, P.K., Anand, S., Da, R.P., 1998. Studies on reduction of manganese dioxide by $(NH_4)2SO_3$ in ammoniacal médium. Hydrometallurgy. 50, 39–49.

Dean, J.A., 1999. Lange's Handbook of Chemistry. New York, N.Y., McGraw-Hill.

Demirbas, A., 1999. Non-isothermal leaching kinetics of braunite in water saturated with sulphur dioxide. Resour. Conserv. Recycl. 26, 35–42.

Deveci, H., Akcil, A., Alp, I., 2003. Parameters for Control and Optimization of Bioleaching of Sulphide Minerals. MS&T, 77–90.

Dew, D.W., Lawson, E.N., Brodhurst, J.L., 1998. The BIOX process for biooxidation of gold-bearing ore or concentrates. D.L. Rawling (Ed.). Biomining Theory, Microbes and Industrial Processes. Springer-Verlag. Berlin, Germany, 45–80.

Elzeky, M., Attia, Y.A., 1995. Effect of bacterial adaptation on kinetics and mechanisms of bioleaching ferrous sulfides. The Chemical Engineering Journal and the Biochemical Engineering Journal, 56(2), B115–B124. https://doi.org/10.1016/0923-0467(94)06086-X

Galen, H., Martinus, H., 1925. The treatment of manganese-silver ores. Hubert work, secretary.

García-Moyano, A., Gonzalez-Toril, E., Moreno-Paz, M., Parro-García, V., Amils, R., 2008. Evaluation of *Leptospirillum* spp. in the Río Tinto, a model of interest to biohydrometallurgy. Hydrometallurgy. 94, 155–161. 10.1016/j.hydromet.2008.05.046.

Greene, A.C., Patel, K.C., Sheehy, A.J., 1997. *Deferribacter thermophilus* gen. nov., a Novel thermophilic manganese and iron reducing bacterium isolated from a petroleum reservoir. Strain, 47(2), 505–509.

Guerinot, M., 1994. Microbial Iron Transport. Annu. Rev. Microbiol., 48, 743–772.

Haines, A.K., 1996. Gold 100. Extracting metallurgy of gold. The South African Institute of Mining and Metallurgy. Johannesburg. 2, 227–233.

Haoran, L., Yali, F., Shouci, L., Zhuwei, D., 2005. Bio-leaching of valuable metals from marine nodules under anaerobic conditions. Minerals Engineering. 18, 1421–1422.

Hernández, F., Ordóñez, I., Robles, J., Gálvez, P., Cisternas, E., 2017. A Methodology for design and operation of heap leaching systems. Mineral processing and extractive metallurgy review. 38. 10.1080/08827508.2017.1281807.

Jensen, A.B., Webb, C., 1995. Treatment of H_2S-containing gases: a review of microbiological alternatives. Enzyme Microbiol. Technol. 17, 2–10.

Jeon, S.-H., Yoo, K., Alorro, R.D., 2017. Separation of Sn, Bi, Cu from Pb-free solder paste by ammonia leaching followed by hydrochloric acid leaching. Hydrometallurgy, 169, 26–30. https://doi.org/10.1016/J.HYDROMET.2016.12.004

Juárez, A., 2004. Biolixiviación de minerales sulfuro-ferroso en jales: de microorganismos involucrados. Universidad de Colima.

Kane, W.S., Cardwell, P.H., 1974. Process for recovering manganese from its ores, US Patent No. 3, 832,165.

Kim, M.-J., Seo, J.-Y., Choi, Y.-S., Kim, G.-H., 2016. Bioleaching of spent Zn–Mn or Ni–Cd batteries by Aspergillus species. Waste Management, 51, 168–173. https://doi.org/10.1016/J.WASMAN.2015.11.001

Lakshmanan, V.I., 1996. Heap and dump bacterial leaching. Biotechnol. and Bioeng. Symp. No. 16, 351–361.

Li, H., 2005. Bio-leaching of valuable metals from marine nodules under anaerobic condition, 18, 1421–1422. https://doi.org/10.1016/j.mineng.2005.04.001

Lin, W., Wu, K., Lao, Z., Hu, W., Lin, B., Li, Y., Hu, J., 2019. Assessment of trace metal contamination and ecological risk in the forest ecosystem of dexing mining area in northeast Jiangxi Province, China. Ecotoxicology and Environmental Safety, 167, 76–82. https://doi.org/10.1016/J.ECOENV.2018.10.001

Liu, R., Chen, J., Zhou, W., Cheng, H., Zhou, H., 2019. Insight to the early-stage adsorption mechanism of moderately thermophilic consortia and intensified bioleaching of chalcopyrite. Biochemical Engineering Journal. https://doi.org/10.1016/J.BEJ.2019.01.009

Ma, L., Wang, X., Tao, J., Feng, X., Zou, K., Xiao, Y., Liu, X., 2017. Bioleaching of the mixed oxide-sulfide copper ore by artificial indigenous and exogenous microbial community. Hydrometallurgy, 169, 41–46. https://doi.org/10.1016/J.HYDROMET.2016.12.007

Medrano-Roldán, H., Galán Wong, L.J., Dávila Flores, R.T., 2013a. Biotecnología de minerales. México DGEST. 1.

Medrano-Roldán., H., Galán-Wong, L.J., Dávila-Flores, R.T., 2013b. Biotecnología de Minerales. Edited. DGEST. México.

Medrano-Roldán., H., J., Losoya-Amaro, M., Dávila-Flores, R.T., 1992. Biotecnología de Minerales. UBAMARI. 25, 4–37.

Morin, D., Pinches, T., Huismanc, J., Friasd, C., Norberge, A., Forssbergf, E., 2006. BioMinE – Integrated project for the development of biotechnology for metal-bearing materials in Europe. Elsevier, Hydrometallurgy, 83(1–4), 69–76.

Natarajan, K.A., Natarajan, K.A., 2018. Bioleaching of Zinc, Nickel, and Cobalt. Biotechnology of Metals, 151–177. https://doi.org/10.1016/B978-0-12-804022-5.00007-4

Padilla, G.A., Cisternas, L.A., Cueto, J.Y., 2008. On the optimization of heap leaching. Minerals Engineering, 21(9), 673–683.

Patente IMPI-SECOFI 950410. Proceso de Eliminación y Regeneración de Cianuro de Sodio en Efluentes Mineros.

Rodriguez, Y., Ballester, A., Blazquez, M.L., Gonzalez, F., Munoz, J.A., 2003. New information on the sphalerite bioleaching mechanism at low and high temperature. Hydrometallurgy. 71, 57–66.

Salazar, M.F., 1999. Biolixiviación de cobre a partir de un concentrado de calcopirita. Tesis de Maestría. Instituto Tecnológico de Durango. Durango, Dgo. México.

Sand, W., Rhode, K., Sobotke, B., Zenneck, C., 1992. Evaluation of *Leptospirillum ferrooxidans* for leaching. Appl. Environ. Microbiol., 58, 85–92.

Schippers, A., Tanne, C., Stummeyer, J., Graupner, T., 2019. Sphalerite bioleaching comparison in shake flasks and percolators. Minerals Engineering, 132, 251–257. https://doi.org/10.1016/J.MINENG.2018.12.007

Shen, L., Li, X., Zhou, Q., Peng, Z., Liu, G., Qi, T., Taskinen, P., 2018. Sustainable and efficient leaching of tungsten in ammoniacal ammonium carbonate solution from the sulfuric acid converted product of scheelite. J. Clean. Prod., 197, 690–698. https://doi.org/10.1016/J.JCLEPRO.2018.06.256

Silverman, M.P., Erlich, H.L., 1994. Microbial leaching from sulphides ores. Adv. Appl. Microbiol., 6, 153–206.

Sugio, T., Tsujita, Y., Hirayama, K., Inagaki, K., Tano, T., 1988. Mechanism of tetravalent manganese reduction with elemental sulfur by *Thiobacillus ferrooxidans*. Agric. Biol. Chem., 52(1), 185–190.

Suzuki, I., 2001. Microbial leaching of metals from sulfide minerals. Biotechnol. Adv., 19(2), 119–132. https://doi.org/https://doi.org/10.1016/S0734-9750(01)00053-2

Syed, S., 2012. Recovery of gold from secondary sources—A review. Hydrometallurgy 115–116, 30–51.

Sylwestrzak, L., 2010. Copper hydrometallurgy processing technologies. Article, Online available at: http://www.thebeijingaxis.com/news detail. php.

Valayapetre, M., Torma, A.E., 1988. Electrochemical recovery of copper from covellite. Metall. 32, 1120–1123.

Wang, J., Huang, Q., Li, T., Xin, B., Chen, S., Guo, X., Li, Y., 2015. Bioleaching mechanism of Zn, Pb, In, Ag, Cd and As from Pb/Zn smelting slag by autotrophic bacteria. J. Environ. Manag., 159, 11–17. https://doi.org/10.1016/J.JENVMAN.2015.05.013

Wang, Y., Li, K., Chen, X., Zhou, H., 2018. Responses of microbial community to pH stress in bioleaching of low grade copper sulfide. Bioresour. Technol., 249, 146–153. https://doi.org/10.1016/J.BIORTECH.2017.10.016

Zachariades, J.C., Fraser, D.M., 1991. Experimental and modelling studies of the kinetics in the leaching of pyrolusite. J. South Afr. Inst. Min. Metall., 91(8), 277–284. https://doi.org/0038-223x

Zhang, L., Qiu, G.-Z., Hu, Y.-H., Sun, X.-J., Li, J.-H. Gu, G.H., 2008. Bioleaching of pyrite by *A. ferrooxidans* and *L. ferriphilum*. Transactions of Nonferrous Metals Society of China 18(6), 1415–1420.

5

Biotechnology for Metal Mechanic Industrial Wastes

Andrea M. Rivas Castillo

Universidad Tecnológica de la Zona Metropolitan del Valle de México, Boulevard, Miguel Hidalgo y Costilla No. 5, Los Héroes de Tizayuza, 43816, Tizayuca, Hidalgo, México
E-mail: a.rivas@utvam.edu.mx

5.1 Introduction

Metal processing and mechanical manufacturing industries produce a significant amount of pollutants that may be released to the environment, representing a serious threaten to be considered. This sector comprises all the industries related to metallic transformation, lamination, or extrusion, and metallic manufacturing is under constant growth due to the boost of sectors such as automotive and aeronautics. Among the diverse types of industries considered as metal mechanic ones, the iron and steel industry is relevant, as it provides an important base for all other industries because its products are used as raw materials for a wide range of industrial processes.

The involved activities in metal mechanic industrial processes may cause air, water, and even soil pollution due to the fumes and dusts produced, and the water discharges or accidental spills that may get dispersed in ecosystems. Even more, liquid and solid residues discharged by these industries can vary in their physical characteristics, organic loads, and the pollutants contained therein, which constitute a challenge for establishing a general treatment procedure or even the most suitable treatment for a specific residue, commonly composed of metals. In several cases, it has been determined that residues recycling and reuse may represent the best choice, and conventional physical or chemical treatment processes are not sufficiently efficient, placing biotechnological approaches into the scope in order to develop more eco-friendly viable processes for the treatment of high metal content residues produced by this industry.

Although the establishment of biotreatment techniques of metal mechanic residues has still a long way to go, and studies on this behalf are relatively scarce. The present chapter addresses the biotreatment approaches of the metallic wastes produced by this industrial sector, regarding heavy metals, slags, metal-containing paints, varnishes, coatings and solvents, chemical treatment and metal finishing residues, cyanide-metal-containing residues, and metallic acidic and nitrate-rich residues, finishing with a relevant metal mechanic industrial case, the iron and steel industry. Thus, the information presented here tries to compile a significant review about the documented studies performed in this field to date, hoping that it will help interested researchers and industrial entities as a general knowledge base to improve biotechnological procedures that may represent promising alternatives to diminish the environmental impact generated by high metal content in mechanic industry wastes.

5.2 Heavy Metals

Metallurgical residues such as dusts, sludges, and liquid and solid wastes produced by diverse ferrous and non-ferrous operations may be considered as a potential source of heavy metals (Sethurajan, 2015). Procedures of mining, electroplating, galvanizing, metal fusion, and mechanic manufacturing, between others, produce residues with elevated amounts of these poisonous elements (Ahluwalia and Goyal, 2007). Specifically, heavy metals like Zn, Pb, and Cr are broadly used for industrial purposes, representing a serious environmental and healthy threaten. High metal content industrial wastes are commonly treated by conventional methods like chemical precipitation, electrochemical treatment, and ion exchange, but these methods may be partially effective and expensive. Diverse biotechnological procedures have been assessed for the biotreatment of high metal content solid residues and wastewater effluents. Plants like *Centaurea virgata*, *Gundelia tournefortii*, *Scariola orientalis*, *Rreseda lutea*, *Noaea mucronate*, and *Eleagnum angustifolia* have been stated as good candidates for heavy metal accumulation (Gupta et al., 2016). Activated sludge processes have been also assessed, and it has been reported that the sludge age may alter each metal removal capability, as micronutrients and macronutrients availability in the activated sludge influences the types of the microorganisms present (Burgess et al., 1999) and, thus, their metal removal selectivity. On the other hand, inactive biomass of microorganisms and plants has been evaluated in heavy metal biosorption processes, as it represents the advantage of not needing growth supporting conditions. It was reported that the biomass of *A. niger, Penicillium chrysogenum, Rhizopus nigricans, Ascophyllum nodosum, Sargassum natans, Chlorella fusca, Oscillatoria anguistissima, Bacillus firmus,* and *Streptomyces* sp. presents elevated metal removal capabilities in the range of 5–641 mg/g for Pb, Zn, Cd, Cr, Cu, and Ni (Ahluwalia and Goyal, 2007).

Metallothioneins and other metal scavenging agents produced by microbial species may represent promising biotechnological approaches for heavy metal removal from industrial residues, and the procedures and advantages have been widely documented, and detailed information can be found elsewhere (Gupta et al., 2016). Currently, technologies for industrial wastewater treatment are developed under the scope of active biofilms supported on an appropriate solid medium, where pollutants removal (i.e., metals) can be performed by microbial accumulation, precipitation, or transformation. Sulfate reducing bacteria from the genera *Desulfovibrio, Desulfotomaculum, Desulfomonas, Thermodesulfobacterium* and *Desulfobulbus* are capable of mediating bioprecipitation processes, as well as *Citrobacter* sp., which is able to produce HPO^{2-} and precipitate metals by this means (Michalak et al., 2013).

Also, many studies have assessed bioleaching processes using mainly the sulfur oxidizing bacteria *A. ferrooxidans* and/or *A. thiooxidans* or the sulfuric acid produced by them, among other microorganisms. Specifically, metallurgical residues were used to evaluate Cd, Cu, and Zn removal contained therein by a bioleaching process, showing that these metals were leached in 50%, 70%, and 60% of the total content, respectively (Sethurajan, 2015). On the other hand, Cu pyrometallurgical slags are highly produced and they present high residual metal content, and they were inappropriately disposed for decades, representing a serious environmental threaten. Although some amount of these slags is currently used in the production of materials used for construction purposes, it is important to develop biotechnological techniques for their treatment. It has been observed that bacterial mediated weathering of these slags by *Pseudomonas aeruginosa* enhances the release of Si, Fe, Cu, Zn, and Pb from the slags, and up to 90% removal of Cu, Zn, and Fe was reached by a bioleaching process using *A. thiooxidans* (Potysz, 2015). Besides, both acidophilic bacteria, *Acidithiobacillus thioparus* and *A. thiooxidans,* have been identified in sludges, having the ability to decrease the pH to about 2, promoting the solubilization of contained metals like Cd, Cr, Cu, Mn, Ni, Pb, and Zn (Blais et al., 1993).

Achievements have been obtained for the biotreatment of heavy metal contaminated industrial streams by applying a biosorbent technique prepared with chitosan spheres coated by acid treated oil palm shell charcoal, which can perform the sorption procedure by ionic interactions and complexation. These spherical particles presented an adsorption capacity value of 154 mg Cr/g, and also the conversion of Cr (VI) to Cr (III) by chitosan was observed, as previously shown in plant tissues and mineral surfaces (Nomanbhay and Palanisamy, 2005). In addition, pellets have been made from bark flour *Pinus ponderosa* pine to evaluate the removal of toxic metals from aqueous solution, being able to remove Cu, Zn, Cd, and Ni, in 57, 53, 50, and 27 mg/g, respectively, demonstrating the potential that natural residues may have as biosorbents for the removal of heavy metals from waste streams (Oh and Tshabalala, 2007).

5.3 Slags

The residues produced by this kind of industrial processes are hard to treat, even by traditional physicochemical procedures, being the best option to minimize their production, which is the case for salty slags, produced as by-products of the secondary Al industry, and that could be considered for the recovery of the Al contained therein (Gil, 2005; Xiao et al., 2005). In a lot of cases, recycling and reuse seem like the options to be considered. That is also the case of smelting copper slag that has been tried to be reused as an aggregate to increase the resistance of concrete (Cendoya, 2009) or the case of asbestos containing residues, as asbestos containing wastes that were thermally treated can be used as filling materials in mortars, preventing its dangerous disposal in landfills (Yvon and Sharrock, 2011). The specific case of Cu pyrometallurgical slags is addressed in the heavy metal section due to its high heavy metal content.

5.4 Metal-containing Paint, Varnish, Coating, and Solvent Residues

Paints are commonly used as industrial coatings, and the paint wastewater produced by their usage is characterized by organic matter, salts, and high suspended solids (Aboulhassan et al., 2006), including heavy metals and recalcitrant organic colorants. Specifically, the volatile organic compounds (VOCs) used in paints and varnishes as solvents, and the metals contained therein may cause significant health impacts, like respiratory, allergic, and immunogenic defects or nerve and kidney damage, respectively (Mendell, 2007). Treatment of this kind of effluents has been assessed using natural adsorbents like clays (Malakootian et al., 2009) and coagulation–flocculation processes (Aboulhassan et al., 2006) On the other hand, biotechnological approaches using plant extracts have also been used for paint wastewater treatment, where it was observed that *Nymphae ampla* extract can considerably reduce Cr (950 mg/L), along with other compounds like chlorides, sulphates, and nitrates (Sharmila et al., 2013).

5.5 Chemical Treatments and Metal Finishing

The procedures used for final metallurgic treatments may vary depending on the final industrial purposes, and a broad range of components may be involved, including acids, bases, heavy metals, cyanides, oils, and fats. Thus, waste streams include metallic liquors and spent baths. Recycling has been the first environmental approach, as it may be possible to save water by reorganizing its use during the diverse processes. However, an interesting method has been patented in which metal-oxidizing bacteria of the genera *Acidithiobacillus* sp., *Leptospirillum* sp., *Sulfobacillus* sp., *Acidiphilium* sp., and *Thermosulfidooxidans* sp. were used for a metal surface treatment, where these bacteria are able to oxidize the metal to perform micro-machining and remove a deformed layer by bioleaching (Ko et al., 2017). Thus, further technological methods may emerge to develop more eco-friendly processes to achieve this kind of industrial needs.

5.6 Cyanide-metal-containing Residues

Although cyanides are naturally encountered in some microorganisms, plants, vegetables, and seeds, their broad usage by metallurgical industries for metal finishing procedures produce an elevated amount of cyanide containing residues, commonly complexed with metals, representing a serious environmental threaten due to their elevated toxicity. Chemical and physical processes available for cyanide degradation are commonly expensive and difficult to operate, so biotechnological approaches could be a suitable option for its removal and have been successfully used in large-scale processes. Some microorganisms can use cyanide as a nitrogen and carbon source and convert it to ammonia and carbonate during their metabolism. In this regard, fungi like *Stemphylium loti*, *Gloeocercospora sorghi*, and *Fusarium* sp. are able to degrade cyanide as well as many bacteria, both under aerobic or anaerobic conditions (Naveen and Shubha, 2011). Specially, the genus *Pseudomonas* sp. has been identified with this capability (Akcil et al., 2003). In this regard, a high cyanide tolerant bacterium, *Pseudomonas pseudoalcaligenes* strain CECT 5344 was able to tolerate a cyanide concentration of 30 mM at alkaline pH, and was able to remove 2 mM of free cyanide after two weeks (Igeño et al., 2018). Has been stated the potential of this to treat cyanide-metal complexes, along with the capability of a *Microbacterium kitamiense* and *Achromobacter* sp. co-culture for this purpose, although this latter joint culture was not able to degrade cyanide in its free form (Igeño et al., 2018).

Pseudomonas pseudoalcaligenes strain W2, which was isolated from a mining effluent, is able to tolerate up to 39 mg/L of cyanide concentration and degrade the 60% of the cyanide contained in a wastewater sample (Tiong et al., 2015). On the other hand, it was reported that some *Klebsiella* sp. strains present cyanide degradation ability, which is the case of *K. pneumoniae* strain coded as ATCC 13883, that is able to transform potassium cyanide in the presence of different metallic ions such as Mg, Ni, Co, Fe, Cr, As, and Zn, in variable concentrations (Avcioglu and Bilkay, 2016).

5.7 Metallic Acidic and Nitrate-rich Residues

The effluents discharged by pickling processes also represent a serious environmental threaten due to their acidic nature, where heavy metals and toxic gases are contained. As conventional physicochemical methods are limited on its efficient and economical feasibility to treat this kind of effluents, biological methods have been assessed. Biological biomass (cow dung, chicken manure, pigeon drops, and sewage) containing a microalgae inoculum seemed a good alternative to treat acidic residues in terms of operational feasibility and metal removal capability (Khawal and Verma, 2015).

This complex industry is tightly linked to each country's economy and to whole world's economy and uses an elevated amount of raw materials, energy, and natural resources, leading to the production of hazardous residues that are discharged to the environment. Although first-world countries accounted for approximately two-thirds of global production, this pattern has shifted to developing countries. The trades involved in this kind of industrial activities impact directly on the prices of final products, as there is a need in the transportation of raw materials from coal and ore-rich producing countries in South America, Africa, and Oceania to major producing areas in Europe, North America, and the Far East, followed by the returning of semi-finished and finished products to the first locations (González and Kamiński, 2011).

Steel plants occupy huge areas comprising different sections like raw material handling, sintering plant, coke oven plant, blast furnace, steel melting shop, oxygen plant, rolling mills, and merchant mills (Das et al., 2018). Environmental releases from these industries account for high metal content residues as well as dust emissions, sulfur dioxides, hydrochloric acid, hydrofluoric acid, polycyclic aromatic hydrocarbons, and persistent organic pollutants (Remus et al., 2013). Unfortunately, scarce research has been made to treat this kind of metallic residues. However, promising alternatives are under development, regarding the usage of microorganisms that may remove metals from solid wastes (Rivas-Castillo et al., 2017, 2018), although further studies will be necessary to assess the capability of these microorganisms for the metal removal contained in solid residues from the iron and steel industry. Besides, an enormous amount of water is used in the processes, generating wastewater and sludges in huge quantities, which contains hazardous substances and chemicals, such as heavy metals (Biswas, 2013). Even more, it is necessary to create water treatment technologies that make possible metal removal and water reuse in order to minimize fresh water consumption and avoid toxic effluent discharge (Das et al., 2018).

Specifically, in the steel making industry electroplating wastewater contains metals like Fe^{2+}, Zn, Ni, and Cr, and in this effluent, Fe^{2+} is oxidized into a metal hydroxide by aeration at neutral pH, being removed afterwards by coagulation settling. Using this method, the metals go into the produced sludge, remaining contained therein. However, the iron-oxidizing bacteria *A. ferrooxidans* has been tested to facilitate Fe^{2+} oxidation under aerobic and acidic conditions, and its utilization may represent an advantage in costs (Kanemori et al., 2017).

5.8 Conclusion

Metal mechanic industry is a very important economic sector, which commonly reflects the prosperity and development of each country. Although companies are getting involved to diminish their environmental impact due to their toxic wastes production, it is imminent to establish more sustainable procedures and treatment technologies to overcome their hazardous discharges to ecosystems, which also represent a serious health threaten due to their elevated toxic compounds, such as metals. In this regard, biotechnological approaches may support the development of more eco-friendly technologies to move forward into the achievement of industrial sustainability. Although there is still a lot to be done in this matter, interesting biotechnological bases for the treatment of high metal content residues produced by the metal mechanic industry have been set up, as addressed in the current chapter. These approaches may reflect promising biotreatment alternatives to be considered for further research in order to scale them into feasible industrial applications.

References

Aboulhassan, M.A., Souabi, S., Yaacoubi, A., Baudu, M., 2006. Improvement of paint effluents coagulation using natural and synthetic coagulant aids. J. Hazard. Mater. 138, 40–45. https://doi.org/10.1016/j.jhazmat.2006.05.040

Ahluwalia, S.S., Goyal, D., 2007. Microbial and plant derived biomass for removal of heavy metals from wastewater. Bioresour. Technol. 98, 2243–2257. https://doi.org/10.1016/j.biortech.2005.12.006

Akcil, A., Karahan, A.G., Ciftci, H., Sagdic, O., 2003. Biological treatment of cyanide by natural isolated bacteria (*Pseudomonas* sp.). Miner. Eng. 16, 643–649. https://doi.org/10.1016/S0892-6875(03)00101-8

Avcioglu, N.H., Bilkay, I.S., 2016. Biological treatment of cyanide by using *Klebsiella pneumoniae* species. Food Technol. Biotechnol. 54(4), 450–454. https://doi.org/10.17113/ftb.54.04.16.4518

Biswas, J., 2013. Evaluation of various method and efficiencies for treatment of effluent from iron and steel industry—a review. Int. J. Mech. Eng. & Rob. Res. 2(3), 67–73.

Blais, J.F., Tyagi, R.D., Auclair, J.C., 1993. Bioleaching of metals from sewage sludge: Microorganisms and growth kinetics. Water Res. 27, 101–110. https://doi.org/10.1016/0043-1354(93)90200-2

Burgess, J., Quarmby, J., Stephenson, T., 1999. Role of micronutrients in activated sludge-based biotreatment of industrial effluents. Biotechnol. Adv. 17, 49–70. https://doi.org/10.1016/S0734-9750(98)00016-0

Cendoya, P., 2009. Efecto en la resistencia de las escorias de fundición de cobre como agregado fino en el comportamiento resistente del hormigón. Ingeniare Rev. Chil. Ing. 17, 85–94. https://doi.org/10.4067/S0718-33052009000100009

Das, P., Mondal, G.C., Singh, S., Singh, A.K., Prasad, B., Singh, K.K., 2018. Effluent treatment technologies in the iron and steel industry – A state of the art review. Water Environ. Res. 90, 395–408. https://doi.org/10.2175/106143017X15131012152951

Gil, A., 2005. Management of the salt cake from secondary aluminum fusion processes. https://doi.org/10.1021/IE050835O

González, I.H., Kamiński, J., 2011. The iron and steel industry: a global market perspective. Gospod. Surowcami Miner. 27, 5–28.

Gupta, A., Joia, J., Sood, A., Sood, R., Sidhu, C., Kaur, G., 2016. Microbes as potential tool for remediation of heavy metals: A review. J. Microb. Biochem. Technol. 8, 364–372. https://doi.org/10.4172/1948-5948.1000310

Igeño, M.I., Macías, D., Guijo, M.I., Sánchez-Clemente, R., Población, A.G., Merchán, F., Blasco, R., 2018. Bacterial consortiums able to use metal-cyanide complexes as a nitrogen source. Proceedings 2, 1284. https://doi.org/10.3390/proceedings2201284

Kanemori, N., Kato, F., Arai, S., Okunuki, S., Kato, T., Mori, N., Kimura, T., Suzuki, Y., 2017. Biological treatment process of steelmaking wastewater treatment. Futtsu.

Khawal, S., Verma, S., 2015. Comparison of biological and algal treatment of acidic wastewater of steel industry. Int. J. Chem. Stud. 3(4): 33–35.

Ko, T.J., Kim, M.Y., Saragih, A.S., 2017. Method for surface treatment of metals using bacteria. US20170152603A1.

Malakootian, M., Nouri, J., Hossaini, H., 2009. Removal of heavy metals from paint industry's wastewater using Leca as an available adsorbent. Int. J. Environ. Sci. Technol. 6, 183–190. https://doi.org/10.1007/BF03327620

Mendell, M.J., 2007. Indoor residential chemical emissions as risk factors for respiratory and allergic effects in children: A review. Indoor Air 17, 259–277. https://doi.org/10.1111/j.1600-0668.2007.00478.x

Michalak, I., Chojnacka, K., Witek-Krowiak, A., 2013. State of the art for the biosorption process – a review. Appl. Biochem. Biotechnol. 170, 1389–1416. https://doi.org/10.1007/s12010-013-0269-0

Naveen, D., Shubha, D., 2011. Biological treatment of cyanide containing wastewater. Res. J. Chem. Sci. 1(7), 15–21. ISSN 2231-606X.

Nomanbhay, S.M., Palanisamy, K., 2005. Removal of heavy metal from industrial wastewater using chitosan coated oil palm shell charcoal. Electron. J. Biotechnol. 8(1), 43–53. https://doi.org/10.2225/vol8-issue1-fulltext-7

Oh, M., Tshabalala, M., 2007. Pelletized ponderosa pine bark for adsorption of toxic heavy metals from water. BioResources. 2, 66–81. https://doi.org/10.15376/biores.2.1.66-81

Potysz, A., 2015. Copper metallurgical slags: mineralogy, bio/weathering processes and metal bioleaching. Paris.

Remus, R., Aguado-Monsonet, M.A., Roudier, S., Delgado Sancho, L., 2013. Best available techniques (BAT) reference document for iron and steel production. Joint Research Centre of the European Commision, Seville. https://doi.org/10.2791/97469

Rivas-Castillo, A., Orona-Tamayo, D., Gómez-Ramírez, M., Rojas-Avelizapa, N.G., 2017. Diverse molecular resistance mechanisms of *Bacillus megaterium* during metal removal present in a spent catalyst. Biotechnol. Bioprocess Eng. 22, 296–307. https://doi.org/10.1007/s12257-016-0019-6

Rivas-Castillo, A., Gómez-Ramirez, M., Rodríguez-Pozos, I., Rojas-Avelizapa, N.G., 2018. Bioleaching of metals contained in spent catalysts by *Acidithiobacillus thiooxidans* DSM 26636. Int. J. Biotechnol. Bioeng. 12, 5.

Sethurajan, M., 2015. Metallurgical sludges, bio/leaching and heavy metals recovery (Zn, Cu). Paris.

Sharmila, S., Rebecca, L.J., Saduzzaman, M.D., 2013. Effect of plant extracts on the treatment of paint industry effluent. Int. J. Pharm. Bio. Sci., 4(3), B678–B686.

Tiong, B., Bahari, Z.M., Lee, N.S.I.S, Jaafar, J., Ibrahim, Z., Shahir, S., 2015. Cyanide Degradation by *Pseudomonas pseudoalcaligenes* Strain W2 Isolated from Mining Effluent (Penguraian Sianida oleh *Pseudomonas pseudoalcaligenes* Strain W2 Dipencilkan daripada Sisa Air Lombong), Sains Malaysiana., 44(2), 233–238.

Xiao, Y., Reuter, M.A., Boin, U., 2005. Aluminium recycling and environmental issues of salt slag treatment. J. Environ. Sci. Health Part A. 40, 1861–1875. https://doi.org/10.1080/10934520500183824

Yvon, Y., Sharrock, P., 2011. Characterization of thermochemical inactivation of asbestos containing wastes and recycling the mineral residues in cement products. Waste Biomass Valorization. 2, 169–181. https://doi.org/10.1007/s12649-011-9063-9

6

Metal Resistance Genes in Microorganisms, a Biotechnological Approach for Ni-V Removal

Grisel Fierros Romero[1,*], José Rubén Mundo Cabello[1] and Reynaldo C. Pless[2]

[1]Tecnológico de Monterrey, School of Engineering and Sciences,
Campus Querétaro, Avenida Epigmenio González No. 500, San Pablo, 76130, Querétaro, México
[2]Departamento de Biotecnología, Centro de Investigación en Ciencia Aplicada y Tecnología Avanzada del Instituto Politécnico Nacional, Cerro Blanco 141, Colinas del Cimatario, 76090 Santiago de Querétaro, Querétaro, México
E-mail: gfierros@itesm.mx
*Corresponding Author

6.1 Introduction

There is incredible richness all around us, hidden in the depths of the ocean and in the hazardous environment of magma chambers, and right under our eyes, in the puddles left by a rainy day or in our organic trash bag. This abundance is biological as well as genetic and it is found in the microorganisms with whom we share our lives and ecosystems. From a biotechnological standpoint, these genes are a powerful resource; they are not only widely available for investigation, but they can also afford important improvements depending on the area of application, which, in the case of this chapter, is the biological treatment of heavy metals to restore the health of an ecosystem. Some metals are required in metabolic processes of microorganisms; these are indispensable as micronutrients and are generally required in trace concentrations (trace elements) such as Co, Cr, Ni, Fe, Mn, Zn, etc. (Nies, 2000). Some of these metals are involved in oxide

109

reduction processes, stabilizing molecules through electrostatic interactions, as catalysts in enzymatic reactions and regulating the osmotic balance (Bruins et al., 2000). However, other heavy metals, such as cadmium and mercury, have no biological function (Roane and Pepper, 1999). In general, it has been described that the accumulation of metals in the soil reduces the microbial biomass (Brookes and McGRATH, 1984) and the enzymatic activity, causing a decrease in the functional diversity of the ecosystem (Ellis et al., 2001).

6.2 Heavy-metal Resistance

In studies carried out in *Escherichia coli*, the minimum inhibitory concentrations (MICs) of different metal ions were tested, showing that the most toxic is mercury and the least is manganese (Mergeay et al., 1985). So far, there are six known mechanisms in bacteria that give them resistance against toxic metals, including exclusion by a permeability barrier, intracellular or extracellular sequestration (mobilization and immobilization), detoxification through active transport of the metal to the outside of the cell, and enzymes that modify the oxidation state of the elements (Bruins et al., 2000; Nies, 2003).

Among the mechanisms of immobilization are the excretions of siderophores, which are extracellular components of some microorganisms, which sequester and solubilize iron; although, these can also chelate other metal ions with lower affinity, such as Ag^+, Al^{3+}, Cd^{2+}, Co^{2+}, $Cr^{2+,3+}$, Cu^{2+}, Eu^{3+}, Ga^{3+}, Hg^{2+}, Mn^{2+}, Ni^{2+}, Pb^{2+}, Sn^{2+}, Tb^{3+}, Tl^+, and Zn^{2+} (Clancy et al., 2006). Another example is the biosurfactants that also have the capacity to remove EPTs (Maier and Soberón-Chávez, 2000). In relation to biosurfactants, it has been described that rhamnolipids are able to remove metals such as cadmium in different types of soil, with a yield of 80% (Wang and Chen, 2006).

The cation diffusion facilitators (CDF) have affinity for different metals, and three main types of cation ejection systems have been identified (Kolaj-Robin et al., 2015):

(I) The CDF are proteins that are distributed in the three domains of life (bacteria, archaea, and eukarya). They generally transport zinc but can also expel other cations such as cadmium, cobalt, nickel, and even iron (Kolaj-Robin et al., 2015; Sahu et al., 2013). CDF proteins are medium-sized polypeptides (less than 400 residues) that possess six transmembrane segments (STM), which function as homodimers in the inner membrane and expel substrates to the periplasmic space. The transport of ions is associated with a chemosmotic process that involves the exchange with protons in which residues of histidine, aspartate, and glutamate participate. The prototype of this group of transporters is encoded by the *CzcD* gene, identified in a plasmid of *Cupriavidus metallidurans* (Anton et al., 2004).

(II) P-type ATPases (so called because the phosphate released in the hydrolysis of ATP binds covalently to the enzyme) constitute a super family of metal transporters that are energized by the hydrolysis of ATP (Sandrin et al., 2000). Like CDF proteins, type-P ATPases are widely distributed among organisms and their substrates are ions such as H^+, Na^+, K^+, Mg^{2+}, Ca^{2+}, Cu^+, Ag^+, Zn^{2+}, and Cd^{2+}. This type of ATPase is located in the inner membrane and can transport ions to the cellular interior, commonly physiological ions such as Mg^{2+}, or function as ion/cation transporter systems, removing toxic metals to the periplasmic space. These ATPases must associate their function with outer membrane proteins (e.g., porins) to expel the metals to the cell exterior. In bacteria, the most studied type P-type ATPase is the CadA protein encoded by *cadA*, a cadmium resistance gene (Icgen and Yilmaz, 2016). However, *C. metallidurans* is noteworthy for having in its genome several genes for 10 P-type ATPases that, taken together, participate in homeostasis, or in resistance, to cadmium, copper, lead, and zinc. Other bacterial groups such as *Cyanobacteria* sp., *Mycobacteria* sp., or *Rhizobium* sp. also harbor a large number of metal ATPases in their genomes (Nies, 2000; von Rozycki and Nies, 2009).

(III) The third group of proteins involved in the metal homeostasis is formed by the family of transporters (RND), named thus because their members participate in processes of resistance, nodulation, and cell division in different organisms and microorganisms (Zhu et al., 2012). These proteins have only been identified in bacteria, where they constitute a great superfamily. These are proteins of around 1000 residues present in the inner membrane, which expel compounds of diverse nature. RND proteins that participate in the expulsion of metals are commonly associated with a pair of auxiliary polypeptides: a small protein from the outer membrane and a periplasmic protein that binds to the inner and outer membranes (Zhang et al., 2012).

6.3 Nickel Resistance

Nickel is an abundant element; it constitutes about 0.008% of the Earth's crust and 0.01% of igneous rocks. Nickel can exist in oxidation states 0, +2, and +3. In addition to simple compounds or salts, nickel forms a variety of coordination compounds or complexes that may be present in enzymes. Some hydrogenases contain nickel, especially those whose function is to oxidize hydrogen. There are two types of nickel transport pumps identified in bacteria (Mulrooney and Hausinger, 2003). These mechanisms are also capable of transporting Co^{2+}; therefore, they are also called NiCoT systems (Hebbeln and Eitinger, 2004). Nickel entry to the cell occurs through a chemosmotic polypeptide transporter or through one of the ABC (ATP-binding cassette) ATPase transport (Mulrooney and Hausinger, 2003). The ABC nickel transporter genes are *NikA* (a soluble nickel-binding protein), *NikB*, *NikC* (membrane proteins that make up the channel), *NikD*, and *NikE* (which form the internal pump complex ATPase). NiCoT systems are also found in archaea and fungi (Hebbeln and Eitinger, 2004). Permease genes of this type have been identified in *C. metallidurans* (gene *HoxN*) and in *Helicobacter pylori* (permease gene *NixA*) (Mulrooney and Hausinger, 2003). These two genes are capable of mediating nickel transport when expressed in *E. coli* (Wolfram et al., 1995). Another similar transport system that confers resistance to Ni/Co is encoded by the permease gene *yohM* described in *E. coli* (Rodrigue et al., 2005). In addition, polypeptides having eight trans-membrane alpha helices with a HisX4AspHis region have been described in helix 2 of HoxN; these motifs being responsible for nickel binding (Degen and Eitinger, 2002).

The energy-dependent nickel-efflux transport systems belonging to the high-affinity ABC transporters (Nies, 2003) have been described in *C. metallidurans* (Lohmeyer and Friedrich, 1987), *Anabaena cylindrical* (Campbell and Smith, 1986), *Methanobacterium bryantii* (Jarrell and Sprott, 1982), and *Clostridium thermoaceticum* (Lundie et al., 1988). These systems are regulated by operons such as *cnrYXHCBA*, *czcCAB*, and *nccCAB*, which modulate the entry of ions into the cell through active transport (ATPase pump) or diffusion (proton pumps) (Gutiérrez et al., 2009; Icgen and Yilmaz, 2016; Nies, 2003). The *cnrYXHCBA* operon of *C. metallidurans* is the most extensively studied genetic determinant that regulates nickel resistance; it does so up to a nickel concentration of 10 mM. This operon is composed of the structural genes *cnrC* (44-kDa protein), *cnrB* (40 kDa), and *cnrA* (115.5 kDa) and the loci *cnrH* (11.6 kDa), *cnrR* (15.5 kDa), and *cnrY* (Liesegang et al., 1993; Stoppel and Schlegel, 1995). The *czcD* gene, initially identified in a plasmid of *C. metallidurans*, acts as a regulator of the *czcCAB* operon that was initially interpreted as conferring resistance to cadmium, zinc, and cobalt, although it has recently been identified in nickel-resistant (Abou-Shanab et al., 2007; Fierros-Romero et al., 2016; Stoppel and Schlegel, 1995). *Czc* is located in the plasmid pMOL30 of *C. metallidurans* and contains four open reading frames (ORFs), which encode four polypeptides, CzcA,

CzcB, CzcC, and CzcD, with transmembrane domains (Diels et al., 1995; Nies, 1992).

The high-expression operon *nccCBA* and some of its genes (nickel–cobalt–cadmium resistance) of *C. metallidurans* have been identified in other bacteria such as *Bacillus megaterium*, *M. liquefaciens*, *Achromobacter xylosoxidans*, *Sphingobacterium heparinum*, *Burkholderia*, *Comamonas*, *Flavobacteria* sp., and *Arthrobacter* sp. (Brim et al., 1999; Dong et al., 1998; Fierros-Romero et al., 2016, 2017). The *nccCBA* complex has structural similarities with the *czcCBA* operon, which is a proton/cation antiporter (Schmidt and Schlegel, 1994). The *ncc* resistance genes have been localized in the 14.5-kb fragments obtained by the *BamHI* enzyme from plasmids pTOM8, pTOM9, and pGOE2 from *Alcaligenes xylosoxidans*; these fragments have been successfully expressed in *E. coli*. These genes are also found in *C. metallidurans* CH34 in the plasmid pMOL28 in a 163-kb fragment (Schmidt and Schlegel, 1994).

6.4 Vanadium Resistance

Vanadium is a transition element of atomic number 23. It is a soft metal, grayish white, malleable, and ductile. The element is naturally found in minerals and fossil fuel deposits. The most representative minerals are patronite (a complex sulfide), vanadinite $[Pb_5(VO_4)_3Cl]$, and carnotite $[K_2(UO_2)_2(VO_4)_2 \cdot 3H_2O]$. Vanadium occurs in four oxidation states: $+2$, $+3$, $+4$, and $+5$. Among the most common V compounds are vanadium dioxide (VO_2) and some vanadates. The vanadates contain pentavalent vanadium; they are analogous to phosphates. Vanadic acid, like phosphoric acid, occurs in ortho, pyro, and meta forms (H_3VO_4, $H_2V_2O_7$, HVO_3). Unlike the situation with phosphoric acid salts, metavanadates are the most stable and orthovanadates are the least stable. In strongly acidic solutions, dioxovanadium (V) cations (VO_2^+) are present. Under physiological conditions, vanadium (V) occurs predominantly as the vanadate anion ($H_2VO_4^-$) and the vanadyl cation (VO_2^+), though other species of cations (VO_3^+, VO_2^+) and anions (HVO_4^{2-}, $V_4O_{12}^{4-}$, $V_{10}O_{28}^{6-}$, $[(VO)_2(OH)_5]^-$) may also occur. At a pH close to 7, vanadium (III) is found exclusively in the form of the V^{3+} cation and in the cellular environment in the form of complexes. Under the acidic conditions of pH 3.5, the vanadyl ion is very stable, while in basic solutions the main form is the orthovanadate ion (VO_4^{3-}), which is very similar in its geometry to the orthophosphate ion (PO_4^{3-}) (Crans et al., 1989; Rehder, 1991). In some cases, vanadate can replace the phosphate of enzymes causing different metabolic disorders (OKAYAMA, 2012). Vanadium may occur in high concentrations in fossil fuels. The concentration of vanadium in crude oil varies, depending on the source, between 0.02 and 1180 ppm. Environmental pollution by vanadium generally comes from processes related to the petrochemical industry (Top et al., 1994). Some organisms show resistance to vanadium through phosphate transport systems (Mahanty et al., 1991) or through the reduction of vanadium (V) to a lower oxidation state, which is less toxic (Carpentier et al., 2003; Ortiz-Bernad et al., 2004). *Geobacter metallireducens* and *Shewanella oneidensis* can grow with vanadium (V) as the sole electron acceptor during anaerobic respiration (Ortiz-Bernad et al., 2004). In South African gold mines, it has been found that *Enterobacter cloacae* is able to reduce a variety of metals, including vanadium, under aerobic and anaerobic conditions, through an unidentified vanadate-reductase enzyme associated with NADH (van Marwijk et al., 2009).

There are few genetic studies on the resistance of bacteria to vanadium. Alonso et al. (2008) described the isolation from the environment of bacteria such as *Escherichia hermannii* and *E. cloacae*, using high vanadyl sulfate concentrations, and showed the induction of an outer membrane protein of 45 kDa, which may be indicative of the induction of an efflux porin system. An iron-dependent superoxide dismutase (sodB) enzyme (Alonso et al., 2008; Kung et al., 2013) and an efflux pump identified in *P. aeruginosa*, MexGHI-OpmD genes (Vandermeulen et al., 2011), have been reported among the mechanisms of vanadium resistance. Regarding the mechanism of vanadium entry, it has been shown that vanadate can enter the cell via the phosphate transport system, both in erythrocytes (Cantley et al., 1978) and in *Neurospora crassa* (Bowman, 1983). In addition, the *VAN1* and *VAN2* genes have been identified in *Saccharomyces*, which confer vanadium resistance through poorly studied mechanisms. The *VAN2* gene (also known as VRG4) codes for a 39.6-kDa protein with transmembrane domains and deletions of this gene are lethal in *Saccharomyces* (Kanik-Ennulat and Neff, 1990; Poster and Dean, 1996).

6.5 Nickel–vanadium Resistance in *Bacillus megaterium* and *Microbacterium liquefaciens*

Bacillus megaterium is a ubiquitous Gram-positive bacterium which thrives in different environments. It is a common model organism for a variety of cellular processes, including morphology, replication, and sporulation, and it is used industrially in the production of recombinant proteins

(Korneli et al., 2013) and of vitamin B12 (Eppinger et al., 2011). Recently, *B. megaterium* has been characterized by an intrinsically high level of resistance to hostile conditions including 20% of salinity (Pal et al., 2014; Salgaonkar et al., 2013) and exposure to metals (e.g., mercury and nickel) (Arenas-Isaac et al., 2017; Narita et al., 2003; Rajkumar et al., 2013). Because of this capability, *B. megaterium* has been used for the biological treatment of wastes with a high content of metals like nickel, vanadium, rhenium, and platinum (Arenas-Isaac et al., 2017; Eppinger et al., 2011; Liu et al., 2011) and to improve the strength of concrete (Krishnapriya et al., 2015). *Microbacterium* species are usually Gram-positive, aerobic or facultative anaerobic, non-motile, non-spore-forming rods. They grow between 15 and 37°C, in the pH range 5–8. *M. liquefaciens* is ubiquitous and has been isolated from contaminated sites (Abou-Shanab et al., 2007; Lima de Silva et al., 2012), showing an intrinsically high level of resistance to metals including 15 mM Ni, 5 mM As, 2.5 mM Cr, and 0.05 mM Hg (Abou-Shanab et al., 2007). In particular, nickel is one of the most hazardous metal species present in various industrial waste streams, such as the ones from petrochemical processes, which create serious problems in terms of environmental preservation due to their accumulation (Amer, 2002). Nickel has biological activity at nanomolar concentrations in bacteria as an essential component of enzymes, mainly ureases and hydrogenases (Mulrooney and Hausinger, 2003). On the other hand, vanadium is found in large amounts in crude oil (Beolchini et al., 2010). It is poorly understood in terms of its biological metabolism in bacteria and only a few vanadium enzymes have been described, such as the vanadium nitrogenases (Lee et al., 2010).

The *B. megaterium* genome contains predicted open reading frames (ORFs) of 110 genes involved in stress responses: 46 in oxidative stress (11 for protection from reactive oxygen species [ROS], 24 for oxidative stress, 2 for glutathione: non-redox reactions, 6 for redox-dependent regulation, 2 for glutaredoxins, and 1 for the CoA disulfide thiol-disulfide redox system), 26 for osmotic stress, 14 for heat shock, 2 for cold shock, and 1 for detoxification (Pal et al., 2014). Some Ni-V resistance genes including proton efflux/influx pumps: *czcD* cation diffusion facilitators (CDF) carrying cadmium, cobalt, nickel, and iron ions, *nccA*, *hant* (high affinity nickel transporter) (Eppinger et al., 2011), *VAN2*, and *smtAB*, a metallothionein (Pazirandeh et al., 1995); some of these genes were previously identified by a PCR approach in *M. liquefaciens* MNSH2-PHGII-2 (GenBank accession number KJ848325.1) (Fierros-Romero et al., 2015) strain that is able to remove Ni and V from spent petrochemical catalyst (Gómez-Ramírez et al., 2015). Differential gene expression induced by exposure to Ni-V is poorly understood.

The presence of the *czcD* gene in *B. megaterium* and *M. liquefaciens* was previously demonstrated by the sequencing of PCR products obtained with the appropriate pairs of primers (Fierros-Romero et al., 2015). There are some reports about the expression changes in the *czcD* gene in *B. megaterium* and *M. liquefaciens* under nickel exposure by real-time RT-PCR (data not published).

Regarding some results from previous works, we established a model with some transport mechanisms of nickel and vanadium in bacteria (Figure 6.1). The *smtAB* gene codes for a metallothionein protein that binds non-specifically to toxic metals, when the bacterial cell is in the presence of these elements (nickel, silver, mercury, arsenic, lead, cadmium, etc.) (Naz et al., 2005). The mechanisms used by *nccA* and *czcD* transporters involve ATP-dependent proton pumps to eject or internalize metals (Anton et al., 2004; Nies, 2003; Schmidt and Schlegel, 1994). Concerning the transporters encoded by *VAN2*, *MexGHI-OpmD*, and *hant* genes, it has been described that their location is in the membrane (Aendekerk et al., 2005; Liu et al., 2011); however, the transport mechanisms of these proteins are little studied. Once inside the cell, the metal can be accumulated, forming complexes with other proteins such as metallothioneins through their cysteine residues (e.g., smtAB); these complexes decrease the toxicity of some metals so that the cells maintain homeostasis. On the other hand, the metal ion can complex with other molecules by remaining inside and/or forming molecules with lower toxicity through chemical changes and being expelled back into the extracellular space (Moore and Helmann, 2005). As mentioned above,

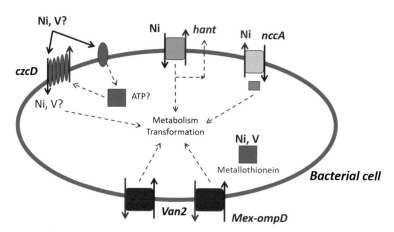

Figure 6.1 Ni-V transport mechanisms in *B. megaterium* and *Microbacterium liquefaciens*.

the transporters are membrane-bound; they internalize or expel the ions as a survival mechanism when exposed to toxic metals. The transductional mechanisms have not yet been elucidated. It is known that once inside the cell, metals can be involved in metabolic processes (those metals that have biological functions), can accumulate forming complexes with proteins such as metallothioneins, or can be transformed into different molecules that are not toxic to the cell, staying within the cell or being expelled via transporters to the extracellular medium (Moore and Helmann, 2005; Nies, 2003; Schmidt et al., 2010).

6.6 Nickel and Vanadium Gene Studies and their Relevance for the Environment

Metal pollution as a result of industrial activity has increased in recent decades due to the inadequate disposal of wastes derived from anthropogenic activities. The chemical industries in our country contribute to this problem by generating a large amount of wastes, including spent catalysts containing metallic elements such as nickel and vanadium. Bacteria exposed to toxic elements (metals) in contaminated sites present adaptations that allow them to survive in these extreme environments. Such modifications may be regulated by different genes that confer resistance and regulate metabolic activities. Knowledge of the molecular mechanisms that regulate these processes can then be used in the development of technologies for the treatment of hazardous wastes such as spent catalysts. Bacteria from Guanajuato mining zones have been used in the biological treatment of industrial wastes containing nickel and vanadium, being able to remove 76.1 ppm Ni and 302.1 ppm V from a 16% Ni-V catalyst from petrochemical wastes (Arenas-Isaac et al., 2017; Fierros-Romero et al., 2016). The study of metal-resistance genes is important because of its potential benefit for the environment and for society. Today, we live in a period of environmental deterioration that requires the creation of new technologies for biological treatment of industrial waste and for bioremediation. The molecular research on heavy metal resistance genes may be relevant for the creation of biomarkers of contamination and for the improvement of biological waste treatment processes, among others.

References

Abou-Shanab, R.A.I., van Berkum, P., Angle, J.S., 2007. Heavy metal resistance and genotypic analysis of metal resistance genes in gram-positive and gram-negative bacteria present in Ni-rich serpentine soil and in the rhizosphere of *Alyssum murale*. Chemosphere. 68, 360–367. https://doi.org/10.1016/j.chemosphere.2006.12.051

Aendekerk, S., Diggle, S.P., Song, Z., Høiby, N., Cornelis, P., Williams, P., Cámara, M., 2005. The MexGHI-OpmD multidrug efflux pump controls growth, antibiotic susceptibility and virulence in *Pseudomonas aeruginosa* via 4-quinolone-dependent cell-to-cell communication. Microbiol. Read. Engl. 151, 1113–1125. https://doi.org/10.1099/mic.0.27631-0

Alonso, F., Ramírez, S., Ancheyta, J., Mavil, M., 2008. Alternativas para la recuperación de metales a partir de catalizadores gastados del hidro-tratamiento de hidrocarburos pesados: un caso de estudio. Rev. Int. Contam. Ambient. 24, 55–69.

Amer, A.M., 2002. Processing of Egyptian boiler-ash for extraction of vanadium and nickel. Waste Manag. 22, 515–520.

Anton, A., Weltrowski, A., Haney, C.J., Franke, S., Grass, G., Rensing, C., Nies, D.H., 2004. Characteristics of zinc transport by two bacterial cation diffusion facilitators from *Ralstonia metallidurans* CH34 and *Escherichia coli*. J. Bacteriol. 186, 7499–7507. https://doi.org/10.1128/JB.186.22.7499-7507.2004

Arenas-Isaac, G., Gómez-Ramírez, M., Montero-Álvarez, L.A., Tobón-Avilés, A., Fierros-Romero, G., Rojas-Avelizapa, N.G., 2017. Novel microorganisms for the treatment of Ni and V as spent catalysts. Indian J. Biotechnol. 16(3), 370–379.

Beolchini, F., Fonti, V., Ferella, F., Vegliò, F., 2010. Metal recovery from spent refinery catalysts by means of biotechnological strategies. J. Hazard. Mater. 178, 529–534. https://doi.org/10.1016/j.jhazmat.2010.01.114

Bowman, B.J., 1983. Vanadate uptake in *Neurospora crassa* occurs via phosphate transport system II. J. Bacteriol. 153, 286–291.

Brim, H., Heyndrickx, M., de Vos, P., Wilmotte, A., Springael, D., Schlegel, H.G., Mergeay, M., 1999. Amplified rDNA restriction analysis and further genotypic characterisation of metal-resistant soil bacteria and related facultative hydrogenotrophs. Syst. Appl. Microbiol. 22, 258–268. https://doi.org/10.1016/S0723-2020(99)80073-3

Brookes, P.C., McGrath, S.P., 1984. Effect of metal toxicity on the size of the soil microbial biomass. J. Soil Sci. 35, 341–346. https://doi.org/10.1111/j.1365-2389.1984.tb00288.x

Bruins, M.R., Kapil, S., Oehme, F.W., 2000. Microbial resistance to metals in the environment. Ecotoxicol. Environ. Saf. 45, 198–207. https://doi.org/10.1006/eesa.1999.1860

Campbell, P.M., Smith, G.D., 1986. Transport and accumulation of nickel ions in the cyanobacterium *Anabaena cylindrica*. Arch. Biochem. Biophys. 244, 470–477.

Cantley, L.C., Resh, M.D., Guidotti, G., 1978. Vanadate inhibits the red cell (Na^+, K^+) ATPase from the cytoplasmic side. Nature 272, 552–554. https://doi.org/10.1038/272552a0

Carpentier, W., Sandra, K., De Smet, I., Brigé, A., De Smet, L., Van Beeumen, J., 2003. Microbial reduction and precipitation of vanadium by *Shewanella oneidensis*. Appl. Environ. Microbiol. 69, 3636–3639. https://doi.org/10.1128/AEM.69.6.3636-3639.2003

Clancy, A., Loar, J.W., Speziali, C.D., Oberg, M., Heinrichs, D.E., Rubens, C.E., 2006. Evidence for siderophore-dependent iron acquisition in group B streptococcus. Mol. Microbiol. 59, 707–721. https://doi.org/10.1111/j.1365-2958.2005.04974.x

Crans, D.C., Bunch, R.L., Theisen, L.A., 1989. Interaction of trace levels of vanadium (IV) and vanadium (V) in biological systems. J. Am. Chem. Soc. 111, 7597–7607. https://doi.org/10.1021/ja00201a049

Degen, O., Eitinger, T., 2002. Substrate specificity of Nickel/Cobalt permeases: Insights from mutants altered in transmembrane Domains I and II. J. Bacteriol. 184, 3569–3577. https://doi.org/10.1128/JB.184.13.3569-3577.2002

Diels, L., Dong, Q., van der Lelie, D., Baeyens, W., Mergeay, M., 1995. The czc operon of *Alcaligenes eutrophus* CH_{34}: from resistance mechanism to the removal of heavy metals. J. Ind. Microbiol. 14, 142–153.

Dong, Q., Springeal, D., Schoeters, J., Nuyts, G., Mergeay, M., Diels, L., 1998. Horizontal transfer of bacterial heavy metal resistance genes and its applications in activated sludge systems. Water Sci. Technol., Microorganisms in Activated Sludge and Biofilm Processes II. 37, 465–468. https://doi.org/10.1016/S0273-1223(98)00147-4

Ellis, R.J., Neish, B., Trett, M.W., Best, J.G., Weightman, A.J., Morgan, P., Fry, J.C., 2001. Comparison of microbial and meiofaunal community analyses for determining impact of heavy metal contamination. J. Microbiol. Methods 45, 171–185.

Eppinger, M., Bunk, B., Johns, M.A., Edirisinghe, J.N., Kutumbaka, K.K., Koenig, S.S.K., Creasy, H.H., Rosovitz, M.J., Riley, D.R., Daugherty, S., Martin, M., Elbourne, L.D.H., Paulsen, I., Biedendieck, R., Braun,

C., Grayburn, S., Dhingra, S., Lukyanchuk, V., Ball, B., Ul-Qamar, R., Seibel, J., Bremer, E., Jahn, D., Ravel, J., Vary, P.S., 2011. Genome sequences of the biotechnologically important *Bacillus megaterium* Strains QM B1551 and DSM319. J. Bacteriol. 193, 4199–4213. https://doi.org/10.1128/JB.00449-11

Fierros-Romero, G., Gomez-Ramirez, M., Arenas-Isaac, G., Pless, R.C., Rojas-Avelizapa, N.G., 2015. Gene identification of *Bacillus megaterium* and *Microbacterium liquefaciens* involved in resistance and metal removal from petrochemical spent catalyst. In Press.

Fierros-Romero, G., Gómez-Ramírez, M., Arenas-Isaac, G.E., Pless, R.C., Rojas-Avelizapa, N.G., 2016. Identification of *Bacillus megaterium* and *Microbacterium liquefaciens* genes involved in metal resistance and metal removal. Can. J. Microbiol. 62, 505–513. https://doi.org/10.1139/cjm-2015-0507

Fierros-Romero, G., Wrosek-Cabrera, J.A., Gómez-Ramírez, M., Pless, R.C., Rivas-Castillo, A.M., Rojas-Avelizapa, N.G., 2017. Expression changes in metal-resistance genes in *Microbacterium liquefaciens* under nickel and vanadium exposure. Curr. Microbiol. 74, 840–847. https://doi.org/10.1007/s00284-017-1252-8

Gómez-Ramírez, M., Montero-Álvarez, L.A., Tobón-Avilés, A., Fierros-Romero, G., Rojas-Avelizapa, N.G., 2015. *Microbacterium oxydans* and *Microbacterium liquefaciens*: a biological alternative for the treatment of Ni-V-containing wastes. J. Environ. Sci. Health Part A Tox. Hazard. Subst. Environ. Eng. 50, 602–610. https://doi.org/10.1080/10934529.2015.994953

Gutiérrez, J.C., Amaro, F., Martín-González, A., 2009. From heavy metal-binders to biosensors: ciliate metallothioneins discussed. BioEssays News Rev. Mol. Cell. Dev. Biol. 31, 805–816. https://doi.org/10.1002/bies.200900011

Hebbeln, P., Eitinger, T., 2004. Heterologous production and characterization of bacterial nickel/cobalt permeases. FEMS Microbiol. Lett. 230, 129–135. https://doi.org/10.1016/S0378-1097(03)00885-1

Icgen, B., Yilmaz, F., 2016. Use of cadA-specific primers and DNA probes as tools to select cadmium biosorbents with potential in remediation strategies. Bull. Environ. Contam. Toxicol. 96, 685–693. https://doi.org/10.1007/s00128-016-1767-x

Jarrell, K.F., Sprott, G.D., 1982. Nickel transport in *Methanobacterium bryantii*. J. Bacteriol. 151, 1195–1203.

Kanik-Ennulat, C., Neff, N., 1990. Vanadate-resistant mutants of *Saccharomyces cerevisiae* show alterations in protein phosphorylation and growth control. Mol. Cell. Biol. 10, 898–909.

Kolaj-Robin, O., Russell, D., Hayes, K.A., Pembroke, J.T., Soulimane, T., 2015. Cation diffusion facilitator family: Structure and function. FEBS Lett. 589, 1283–1295. https://doi.org/10.1016/j.febslet.2015.04.007

Korneli, C., Biedendieck, R., David, F., Jahn, D., Wittmann, C., 2013. High yield production of extracellular recombinant levansucrase by *Bacillus megaterium*. Appl. Microbiol. Biotechnol. 97, 3343–3353. https://doi.org/10.1007/s00253-012-4567-1

Krishnapriya, S., Venkatesh Babu, D.L., Prince Arulraj, G., 2015. Isolation and identification of bacteria to improve the strength of concrete. Microbiol. Res. 174, 48–55. https://doi.org/10.1016/j.micres.2015.03.009

Kung, S.H., Retchless, A.C., Kwan, J.Y., Almeida, R.P.P., 2013. Effects of DNA size on transformation and recombination efficiencies in *Xylella fastidiosa*. Appl. Environ. Microbiol. 79, 1712–1717. https://doi.org/10.1128/AEM.03525-12

Lee, C.C., Hu, Y., Ribbe, M.W., 2010. Vanadium nitrogenase reduces CO. Science 329, 642. https://doi.org/10.1126/science.1191455

Liesegang, H., Lemke, K., Siddiqui, R.A., Schlegel, H.G., 1993. Characterization of the inducible nickel and cobalt resistance determinant cnr from pMOL28 of *Alcaligenes eutrophus* CH34. J. Bacteriol. 175, 767–778. https://doi.org/10.1128/jb.175.3.767-778.1993

Lima de Silva, A.A., de Carvalho, M.A.R., de Souza, S.A.L., Dias, P.M.T., da Silva Filho, R.G., de Meirelles Saramago, C.S., de Melo Bento, C.A., Hofer, E., 2012. Heavy metal tolerance (Cr, Ag and Hg) in bacteria isolated from sewage. Braz. J. Microbiol. Publ. Braz. Soc. Microbiol. 43, 1620–1631. https://doi.org/10.1590/S1517-838220120004000047

Liu, L., Li, Y., Zhang, J., Zou, W., Zhou, Z., Liu, J., Li, X., Wang, L., Chen, J., 2011. Complete genome sequence of the industrial strain *Bacillus megaterium* WSH-002. J. Bacteriol. 193, 6389–6390. https://doi.org/10.1128/JB.06066-11

Lohmeyer, M., Friedrich, C.G., 1987. Nickel transport in *Alcaligenes eutrophus*. Arch. Microbiol. 149, 130–135. https://doi.org/10.1007/BF00425078

Lundie, L.L., Yang, H.C., Heinonen, J.K., Dean, S.I., Drake, H.L., 1988. Energy-dependent, high-affinity transport of nickel by the acetogen *Clostridium thermoaceticum*. J. Bacteriol. 170, 5705–5708.

Mahanty, S.K., Khaware, R., Ansari, S., Gupta, P., Prasad, R., 1991. Vanadate-resistant mutants of *Candida albicans* show alterations in phosphate uptake. FEMS Microbiol. Lett. 68, 163–166. https://doi.org/10.1111/j.1574-6968.1991.tb04590.x

Maier, R.M., Soberón-Chávez, G., 2000. *Pseudomonas aeruginosa* rhamnolipids: Biosynthesis and potential applications. Appl. Microbiol. Biotechnol. 54, 625–633. https://doi.org/10.1007/s002530000443

Mergeay, M., Nies, D., Schlegel, H.G., Gerits, J., Charles, P., Van Gijsegem, F., 1985. *Alcaligenes eutrophus* CH34 is a facultative chemolithotroph with plasmid-bound resistance to heavy metals. J. Bacteriol. 162, 328–334.

Moore, C.M., Helmann, J.D., 2005. Metal ion homeostasis in *Bacillus subtilis*. Curr. Opin. Microbiol. 8, 188–195. https://doi.org/10.1016/j.mib.2005.02.007

Mulrooney, S.B., Hausinger, R.P., 2003. Nickel uptake and utilization by microorganisms. FEMS Microbiol. Rev. 27, 239–261. https://doi.org/10.1016/S0168-6445(03)00042-1

Narita, M., Chiba, K., Nishizawa, H., Ishii, H., Huang, C.-C., Kawabata, Z., Silver, S., Endo, G., 2003. Diversity of mercury resistance determinants among *Bacillus* strains isolated from sediment of Minamata Bay. FEMS Microbiol. Lett. 223, 73–82. https://doi.org/10.1016/S0378-1097(03)00325-2

Naz, N., Young, H.K., Ahmed, N., Gadd, G.M., 2005. Cadmium accumulation and DNA homology with metal resistance genes in sulfate-reducing bacteria. Appl. Environ. Microbiol. 71, 4610–4618. https://doi.org/10.1128/AEM.71.8.4610-4618.2005

Nies, D.H., 1992. Resistance to cadmium, cobalt, zinc, and nickel in microbes. Plasmid. 27, 17–28.

Nies, D.H., 2000. Heavy metal-resistant bacteria as extremophiles: Molecular physiology and biotechnological use of *Ralstonia* sp. CH34. Extrem. Life Extreme Cond. 4, 77–82.

Nies, D.H., 2003. Efflux-mediated heavy metal resistance in prokaryotes. FEMS Microbiol. Rev. 27, 313–339. https://doi.org/10.1016/S0168-6445(03)00048-2

Okayama, H., 2012. Functional cDNA expression cloning: Pushing it to the limit. Proc. Jpn. Acad. Ser. B Phys. Biol. Sci. 88, 102–119. https://doi.org/10.2183/pjab.88.102

Ortiz-Bernad, I., Anderson, R.T., Vrionis, H.A., Lovley, D.R., 2004. Vanadium respiration by *Geobacter metallireducens*: Novel strategy for in situ removal of vanadium from groundwater. Appl. Environ. Microbiol. 70, 3091–3095.

Pal, K.K., Dey, R., Sherathia, D., Vanpariya, S., Patel, I., Dalsania, T., Savsani, K., Sukhadiya, B., Mandaliya, M., Thomas, M., Ghorai, S., Rupapara, R., Rawal, P., Shah, A., Bhayani, S., 2014. Draft genome sequence of a moderately halophilic *Bacillus megaterium* strain, MSP20.1, Isolated from a saltern of the Little Rann of Kutch, India. Genome Announc. 2. https://doi.org/10.1128/genomeA.01134-13

Pazirandeh, M., Chrisey, L.A., Mauro, J.M., Campbell, J.R., Gaber, B.P., 1995. Expression of the *Neurospora crassa* metallothionein gene in *Escherichia coli* and its effect on heavy-metal uptake. Appl. Microbiol. Biotechnol. 43, 1112–1117.

Poster, J.B., Dean, N., 1996. The yeast VRG4 gene is required for normal Golgi functions and defines a new family of related genes. J. Biol. Chem. 271, 3837–3845.

Rajkumar, M., Ma, Y., Freitas, H., 2013. Improvement of Ni phytostabilization by inoculation of Ni resistant *Bacillus megaterium* SR28C. J. Environ. Manage. 128, 973–980. https://doi.org/10.1016/j.jenvman.2013.07.001

Rehder, D., 1991. The bioinorganic chemistry of vanadium. Angew. Chem. Int. Ed. Engl. 30, 148–167. https://doi.org/10.1002/anie.199101481

Roane, T.M., Pepper, I.L., 1999. Microbial responses to environmentally toxic cadmium. Microb. Ecol. 38, 358–364.

Rodrigue, A., Effantin, G., Mandrand-Berthelot, M.A., 2005. Identification of *rcnA (yohM)*, a nickel and cobalt resistance gene in *Escherichia coli*. J. Bacteriol. 187, 2912–2916. https://doi.org/10.1128/JB.187.8.2912-2916.2005

Sahu, K.K., Agrawal, A., Mishra, D., 2013. Hazardous waste to materials: Recovery of molybdenum and vanadium from acidic leach liquor of spent hydroprocessing catalyst using alamine 308. J. Environ. Manage. 125, 68–73. https://doi.org/10.1016/j.jenvman.2013.03.032

Salgaonkar, B.B., Mani, K., Braganca, J.M., 2013. Characterization of polyhydroxyalkanoates accumulated by a moderately halophilic salt pan isolate *Bacillus megaterium* strain H16. J. Appl. Microbiol. 114, 1347–1356. https://doi.org/10.1111/jam.12135

Sandrin, T.R., Chech, A.M., Maier, R.M., 2000. A rhamnolipid biosurfactant reduces cadmium toxicity during naphthalene biodegradation. Appl. Environ. Microbiol. 66, 4585–4588.

Schmidt, A., Hagen, M., Schütze, E., Schmidt, A., Kothe, E., 2010. *In silico* prediction of potential metallothioneins and metallohistins in actinobacteria. J. Basic Microbiol. 50, 562–569. https://doi.org/10.1002/jobm. 201000055

Schmidt, T., Schlegel, H.G., 1994. Combined nickel-cobalt-cadmium resistance encoded by the ncc locus of *Alcaligenes xylosoxidans* 31A. J. Bacteriol. 176, 7045–7054.

Stoppel, R., Schlegel, H.G., 1995. Nickel-resistant bacteria from anthropogenically nickel-polluted and naturally nickel-percolated ecosystems. Appl. Environ. Microbiol. 61, 2276–2285.

Top, E., De Smet, I., Verstraete, W., Dijkmans, R., Mergeay, M., 1994. Exogenous isolation of mobilizing plasmids from polluted soils and sludges. Appl. Environ. Microbiol. 60, 831–839.

van Marwijk, J., Opperman, D.J., Piater, L.A., van Heerden, E., 2009. Reduction of vanadium (V) by *Enterobacter cloacae* EV-SA01 isolated from a South African deep gold mine. Biotechnol. Lett. 31, 845–849. https://doi.org/10.1007/s10529-009-9946-z

Vandermeulen, G., Marie, C., Scherman, D., Préat, V., 2011. New generation of plasmid backbones devoid of antibiotic resistance marker for gene therapy trials. Mol. Ther. J. Am. Soc. Gene Ther. 19, 1942–1949. https://doi.org/10.1038/mt.2011.182

von Rozycki, T., Nies, D.H., 2009. *Cupriavidus metallidurans*: evolution of a metal-resistant bacterium. Antonie Van Leeuwenhoek 96, 115–139. https://doi.org/10.1007/s10482-008-9284-5

Wang, J., Chen, C., 2006. Biosorption of heavy metals by *Saccharomyces cerevisiae*: A review. Biotechnol. Adv. 24, 427–451. https://doi.org/10.1016/j.biotechadv.2006.03.001

Wolfram, L., Friedrich, B., Eitinger, T., 1995. The *Alcaligenes eutrophus* protein HoxN mediates nickel transport in Escherichia coli. J. Bacteriol. 177, 1840–1843.

Zhang, W., Huang, Z., He, L., Sheng, X., 2012. Assessment of bacterial communities and characterization of lead-resistant bacteria in the rhizosphere soils of metal-tolerant *Chenopodium ambrosioides* grown on lead-zinc mine tailings. Chemosphere 87, 1171–1178. https://doi.org/10.1016/j.chemosphere.2012.02.036

Zhu, H., Guo, J., Chen, M., Feng, G., Yao, Q., 2012. *Burkholderia dabaoshanensis* sp. nov., a heavy-metal-tolerant bacteria isolated from Dabaoshan mining area soil in China. PloS One 7, e50225. https://doi.org/10.1371/journal.pone.0050225

7

Bioweathering of Heap Rock Material by Basidiomycetes

Erika Kothe[*], **Julia Kirtzel and Katrin Krause**

Friedrich Schiller University, Institute of Microbiology, Neugasse 25,
D 07743 Jena, Germany
E-mail: erika.kothe@uni-jena.de
*Corresponding Author

The weathering of heap rock material and minerals therein is influenced by bioweathering. The impact of fungi, specifically basidiomycetes, is still not well understood. Here, the potential of the widespread basidiomycete *Schizophyllum commune* to decay black slate, a rock rich in organic carbon, was analyzed to determine the biomolecular processes and the potential impacts on the environment. Biomechanical mechanisms such as a disruption of rocks due to hyphal penetration via a hyphal formation and widening of cracks, production of extracellular mucilaginous material such as schizophyllan affecting physical forces during wetting and drying, and the secretion of enzymes, siderophores, or organic and inorganic acids is reviewed. Signaling pathways and genes involved in black slate bioweathering were addressed with a transcriptome analysis of the wood-degrading basidiomycete *S. commune*. Shotgun proteomics confirmed the relevance of wood-degrading enzymes in metal release during rock weathering. Emphasis is put on black slate weathering that has been mined for uranium in Ronneburg, Thuringia. This allows to address the waste rock material and processes prevailing in heap material. Colonization of rocks, especially those from former mining areas, by basidiomycetes might lead to serious contaminations with (heavy) metals and thus provoke problems for the environment and human health.

7.1 Introduction

Abiotic weathering of rocks and minerals contributes significantly to biogeochemical cycling of elements and it is well recognized that micro-organisms can influence and accelerate weathering processes. During the last century, microbially induced bioweathering of rocks and their constituents gained considerable attention (Gadd, 2007). Most studies focused on the role of bacteria and their bioweathering strategies as well as their positive and negative impacts on the environment. The role of fungi has largely been neglected, but, more recently, fungal bioweathering attracted notice to biogeochemical research. However, very few investigations deal with basidiomycetes and their influence on the biodeterioration of rocks and minerals. Although previous studies discovered that basidiomycetes have an impact on rocks rich in organic carbon (Hofrichter et al., 1999; Wengel et al., 2006), the mechanisms involved in bioweathering are still unclear.

To evaluate bioweathering on a biomolecular level, fungal growth with black slate was addressed. Black slates are fine-grained sedimentary, low-grade metamorphic rocks rich in sulfides, such as pyrite, metals, and organic matter. Unweathered, dark-colored black slate contains about 6 wt.% total organic carbon, whereas strongly weathered, light-colored black slate contains about 0.2–2 wt.% total organic carbon (Fischer et al., 2009). The oxidation of organic matter from black slate leads to the release of CO_2 and thereby significantly contributes to the global carbon cycle (Jaffe et al., 2002).

In the former mine Morassina, located in the Thuringian Forrest in Schmiedefeld, Germany, black slate was intensively mined between 1683 and 1863. Stratigraphically, black slate from the Thuringian Slate Mountains can be attributed to Silurian and Lower Devonian ages and is present in 80 cm to 3 m large seams belonging to graptolite shale series. The occurring alum shale was extracted in underground mines for its sulfur, alum, and vitriol contents in three working horizons.

The filamentous fungus *S. commune* is a widely distributed basidiomycete which can be found on all continents except Antarctica. It forms fan-shaped, grey fruiting bodies which are 1–4 cm wide and can be mainly found on fallen hardwoods (Raper, 1966). The fungus was intensively studied regarding its genetics, mating system, morphogenesis, and physiology. *S. commune* can be grown under laboratory conditions and complete its life cycle within

10–14 days. The genome sequence revealed a diverse set of enzymes involved in, e.g., lignocellulose degradation (Ohm et al., 2010).

7.2 Mechanisms of Bioweathering

Rocks and their mineral constituents can be weathered by physical, chemical, and biological factors, in which the relative importance of each factor depends on environmental conditions, the rock type, and its preservation state (Papida et al., 2000). The mechanisms can be classified into biomechanical and biochemical processes (Lee and Parsons, 1999). Both processes are closely linked since biomechanical weathering increases the surface area of the rock or mineral, which makes it more susceptible to biochemical processes. However, the chemical action of organisms is considered to be a more important process than mechanical degradation (Gadd, 2007).

Biomechanical weathering by fungi can be divided into direct and indirect mechanisms. Direct biodeterioration is facilitated by the hyphal turgor pressure which allows fungi to penetrate rocks, grain boundaries, cleavages, and cracks. Indirect biomechanical weathering occurs with the help of extracellular mucilaginous material (ECMM) which facilitates biofilm formation and attachment to solid surfaces. Swelling and shrinking effects of the biofilm and ECMM lead to mechanical pressure to the mineral lattice and can cause its erosion or abrasion (Warscheid and Krumbein, 1994). In addition, biofilms control the pH and diffusion of ions or chelators and may thereby greatly increase bioweathering. The resulting damage of the rock due to direct and indirect biomechanical processes causes its fractionation and splitting and makes it amenable to biochemical deterioration.

Biochemical weathering includes fungal leaching processes like the solubilization of solid metal compounds and can be achieved through acidolysis, complexolysis, redoxolysis, and the accumulation of metals into the mycelium (Burgstaller and Schinner, 1993). Acidolysis leaches metals from solids by protonation, complexolysis solubilizes metal ions through the formation of ligands, and redoxolysis releases metals from minerals through oxidation and reduction reactions. It is assumed that acidolysis and complexolysis are the primary fungal mechanisms in mineral dissolution. Several processes including proton efflux, the absorption of nutrients in exchange for protons, carbonic acid formation as result of respiratory CO_2 production, as well as the secretion of siderophores, amino acids, and phenolic compounds are involved. One important fungal leaching mechanism comprises the secretion of organic acids that act as strong chelators or decrease pH by providing a source of protons, and thus aggressively attack mineral surfaces (Gadd, 1999). Organic acids can additionally contribute to redoxolysis because they can be a source of electrons which reduce metals to lower oxidation states. Especially for white-rot fungi, redoxolysis could be of great significance since those fungi secrete a variety of enzymes with high redox potentials (Martínez et al., 2005).

Fungal biomass can also function as a sink for mobilized metal cations by metal biosorption to biomass (e.g., cell walls, pigments, and ECMM), intracellular accumulation and sequestration as well as precipitation of metal-containing biominerals onto hyphae (Burford et al., 2003). These mechanisms immobilize metals and reduce the free metal activity in the environment, thus shifting the equilibrium to release more metals (Burford et al., 2003). Biochemical weathering of rocks changes the microtopography of minerals through etching and pitting, and the diagenesis or even complete dissolution of mineral grains (Ehrlich, 1998).

The different mechanisms by which *S. commune* achieves the bioweathering of black slate and their effects are represented schematically in Figure 7.1.

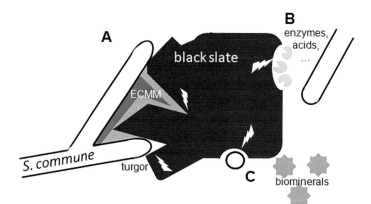

Figure 7.1   Proposed scheme of mechanical and chemical bioweathering mechanisms for *S. commune*. Biomechanical weathering through extracellular mucilaginous material (ECMM; schizophyllan) and turgor pressure causes and widens cracks in black slate (A). Exoenzymes as well as acidification by excretion of organic and inorganic acids cause a biochemical degradation and dissolution of black slate accompanied by metal release (B). The mobilization of metals can lead to formation of biominerals, while acidification and enzymatic attack are involved in formation of etch pits (C).

7.2.1 Biomechanical Weathering

The investigated biomechanical mechanisms showed a close interaction of *S. commune* with the rock material. The fungus either enclosed ground black slate particles in liquid culture or covered pieces of the rock on agar plates with a dense hyphal mat. In liquid culture, SEM analyses showed that hyphae penetrated pre-existing fissures in the rock grains and probably widened them. Furthermore, hyphal tips grew out in the midst of a grain suggesting that hyphae also formed new cracks. It seems likely that *S. commune* used the rough surface of rock particles for a better attachment so that hyphal tips could use depressions as an initial point for the formation of new cracks. This effective direct biomechanical weathering is based on two reasons: the cytoskeleton and an osmotically generated turgor pressure (Money et al., 2004). Although biomechanical processes are often considered to be less effective than biochemical processes, especially invading fungi, like the white-rot fungus *S. commune*, have high turgor pressures and could thus intensively deteriorate rocks by exerting physical forces (Money et al., 2004). As a result, rock grains were crushed into smaller pieces and (mineral) constituents were detached.

Shrinking and swelling of ECMM exert mechanical stress to the rock and mineral structure and alter the pore size distribution, moisture circulation patterns, and temperature response. In nature, ECMM promotes biofilm formation and thus leads to an additional attack by bacteria (Gadd, 2007). Growth of biofilms into cracks further minimizes the surface area towards the soil and prevents rapid diffusion of exudates into the soil solution. A dense network of ECMM could be identified as schizophyllan, a non-ionic, water-soluble homoglucan consisting of a backbone chain of 1,3-β-D-glucopyranose units linked with single 1,6-bounded β-D-glucopyranose (Burford et al., 2003; Freihorst et al., 2016). Schizophyllan, with a molecular weight varying between 6106 and 12,106 g/mol, dissolves in water as a triple helix which results in its well-known viscosifying properties in aqueous solutions (Freihorst et al., 2016).

Hyphal mucilage was shown to contain enzymes like laccases as well as acidic and metal-chelating compounds (Burford et al., 2003). They could possibly react with black slate for the release of metals and lead to a depletion of schizophyllan. This reasoning was further reinforced by the fact that microarray data showed up to 13-fold upregulation of several glucans, including a 1,3-β-glucan (schizophyllan, 2.2-fold upregulation), in the presence of black slate. Thus, an increased production of ECMM in the presence of the rock and its following decay seemed likely. In addition, threads of mucilage were found to cover hyphae and connect them to black slate grains. These connections could enhance the fungal attachment to the rock (Gadd, 2007) and support its physical and chemical deterioration.

7.2.2 Extracellular Enzymes

The group of enzymes specifically involved in redox reactions at stone surfaces are derived from the pathways of organic carbon degradation by fungi. Enzymes potentially involved in lignin catabolism are commonly classified as FOLymes (Fungal Oxidative Lignin enzymes). From the genome of *S. commune*, 16 FOLyme genes and 11 genes distantly related to FOLymes were identified (Ohm et al., 2010). They include genes for one cellobiose dehydrogenase, two laccases, and 13 lignin-degrading auxiliary enzymes but lack genes encoding peroxidases. With a remarkable number of enzymes identified by the CAZy database (Carbohydrate-Active Enzymes), including candidates of 240 glycoside hydrolases, 75 glycosyl transferases, 16 polysaccharide lyases, and 30 carbohydrate esterases, *S. commune* is also well equipped to degrade pectin, cellulose, and hemicellulose (Ohm et al., 2010).

Multicopper oxidases are a family of enzymes comprising laccases (EC 1.10.3.2), ascorbate oxidases (EC 1.10.3.3), and ceruloplasmins (ferroxidases, EC 1.16.3.1) which couple the oxidation of a substrate with a four-electron reduction of O_2 to $2\ H_2O$. Fungal laccases (p-diphenol:dioxygen oxidoreductases) are the most important group of multicopper oxidases with respect to their number and extent of characterization. They oxidize a variety of aromatic hydrogen donors by catalyzing the removal of an electron and a proton from phenolic hydroxyl or aromatic amino groups which results in the formation of free phenoxy radicals or amino radicals, respectively (Leonowicz et al., 2001). In the genome of *S. commune*, two laccase and four laccase-like multicopper oxidase genes (*lcc1* and *lcc2*) (*mco1*, *mco2*, *mco3*, and *mco4*) were identified (Ohm et al., 2010). Besides delignification, fungal laccases are involved in fruiting body development, pigmentation, pathogenesis, coal degradation, as well as in resistance against oxidative stress, and they are used in soil bioremediation for the degradation of polycyclic aromatic hydrocarbons and xenobiotics.

7.2.3 Acidification

Another important process in biochemical weathering is the secretion of organic and inorganic acids. They provide both a source of protons and metal-chelating anions to solubilize minerals and chelate metal cations, respectively, thereby leading to a solubilization of rocks and influencing the mobilization of metals (Gadd, 1999). By providing an additional source of protons, acidolysis provoked through organic acid has often been reported to be the main mechanism in mineral dissolution (Fomina et al., 2005). However, the secretion of large amounts of organic acids with chelating properties resulted in a switch from proton-promoted to ligand-promoted dissolution which finally became more efficient than acidolysis (Fomina et al., 2005).

When *S. commune* was grown with black slate, microarray data showed that many enzymes within the citric acid cycle were upregulated (Kirtzel et al., 2017). D-aspartate oxidase, as one of these enzymes, leads to the generation of hydrogen peroxide which in turn was shown to attack, oxidize, and dissolve minerals as well as metal ions from ore surfaces (Javadi Nooshabadi and Hanumantha Rao, 2014). Furthermore, respiratory CO_2 is produced within the citric acid cycle which could cause an additional acidification by the formation of carbonic acid. Black slate also induced the production of arginase (5.4-fold) and urease (4.3-fold) which might participate in ammonia and carbonic acid formation, thereby leading to alkaline or acidic conditions known to dissolve minerals (Kirtzel et al., 2017). Moreover, many organic acids such as citric, succinic, fumaric, and malic acid are produced in the citric acid cycle and genes potentially involved in their generation were influenced by black slate (Kirtzel et al., 2017). HPLC analysis confirmed this assumption and revealed the secretion of additional organic acids such as gluconic, ascorbic, and maleic acid. Other products of the citric acid cycle like oxaloacetate and fumarate are precursors for the production of oxalic as well as epoxysuccinic and tartaric acid. Black slate and released metals were shown to interact with all organic acids. Ascorbic acid has been shown to accelerate the dissolution of quartz, the most frequent mineral in black slate, whereas tartaric acid is a strong complexing agent and heavily increased the weathering of silicate minerals (Huang and Keller, 1972). Therefore, both acids probably contributed to the dissolution of the three identified minerals in black slate: quartz, kaolinite, and muscovite. Oxalic acid, which was secreted with highest concentrations by *S. commune*, is also a strong solubilizing agent of silicate minerals and thus might participate in the dissolution of black slate. In addition, oxalic acid is a very effective chelator of Fe and Al cations. The presence of both elements in black slate (Kirtzel et al., 2018) underpins the role of this acid in biodeterioration.

7.2.4 Metal Mobilization

Besides Fe, siderophores are known to mobilize, albeit with lower efficiencies, several other metals such as Al, Cd, Co, Cr, Cu, Mn, Ni, Pb, and Zn. All these metals were detected in black slate, and, except Cr, all metals were bioavailable (Kirtzel et al., 2019). During chelation, siderophores do not only promote the dissolution of minerals and rocks but also supply microorganisms or plants with these metals (Ahmed and Holmström, 2014). Taking Fe as example, chelated Fe^{3+} can be reduced by a ferric reductase either before or after transport into the cell (Schröder et al., 2003). Evaluation of microarray data revealed an upregulation of several ferric reductases during black slate treatment and thus indirectly confirmed an enhanced Fe accumulation via siderophores.

7.3 Cellular Responses to Growth on Black Shale

Biomolecular mechanisms involved in rock degradation were examined to identify participating genes and proteins, black slate induced changes in the transcriptome and secretome were evaluated using microarray data and shotgun proteomics, respectively.

The analysis of the proteome showed the production of well over 100 proteins (Kirtzel et al., 2018). For 12 proteins, a consistent induction by incubation with black slate was shown (Kirtzel et al., 2018).

Another biodegradation mechanism comprises the leaching and bioaccumulation of metals from rocks. To evaluate this for *S. commune*, intracellular staining of specific metals was performed and showed a sequestration of Cd, Fe, Pb, and Zn in different cell organelles (Kirtzel et al., 2019). Additionally, ICP-MS (inductively coupled plasma mass spectrometry) measurements revealed the accumulation of further metals into biomass of *S. commune* (Kirtzel et al., 2019). It is known that metal accumulation can reduce the free metal activity in the environment and in turn cause the release of more metals (Gadd, 2007).

A transcriptome analysis was performed on 13.181 predicted genes of *S. commune* $12-43\times4-39$ (Erdmann et al., 2012) and revealed an upregulation of 362 genes (2.7%) and a downregulation of 400 genes (3.0%) during growth with black slate (Kirtzel et al., 2017). Most upregulated genes belonged to oxidoreductase and hydrolase families and secretome analysis reinforced these findings (Kirtzel et al., 2018). A comparison of transcripts and secreted proteins showed consistent regulations of oxidoreductases, such as multicopper oxidases (proteomics) and laccases (microarray), and hydrolases such as glycoside hydrolase families (GH5, GH43, and GH61) and peptidases (peptidase A1, S8, and S53) during growth on black slate (Kirtzel et al., 2017, 2018). The organic matter fraction of black slate, mainly consisting of aromatic hydrocarbons, could thus be attacked by ligninolytic enzymes such as laccases (Seifert et al., 2011).

Besides enzymes responsible for the degradation of organic matter, *S. commune* produced proteins associated with interaction with inorganic matter. They include phosphoadenosine phosphosulfate reductase (7.7-fold) as well as alkaline phosphatase (2.1-fold) and are known to inter-act with minerals and contribute to a cycling of sulfur or phosphorous (Kirtzel et al., 2017). In addition, the transcriptome showed upregulation of hydrophobins (e.g., beta-1,6-N-acetylglucosaminyl transferase, 21-fold) and several lectins (up to 18-fold) (Kirtzel et al., 2017). Lectins are sugar-binding proteins which could be used by fungi for an enhanced adhesion and sub-sequent invasion of trees. Similarly, hydrophobins, which are cysteine-rich hydrophobic proteins, can mediate the attachment of hyphae to hydrophobic surfaces (Wösten, 2001). Thus, lectins and hydrophobins could improve the fungal attachment to rock surfaces and hence promote its growth on black slate.

Analysis of genes and proteins provided a comprehensive picture of biomolecular mechanisms during black slate decomposition and demon-strated a variety of genes and proteins with various functions. This empha-sizes all three, the involvement of lignocellulolytic enzymes in black slate degradation as well as a wide biochemical potential and metal detoxification mechanisms (see Figure 7.2) of *S. commune*.

7.3.1 Extracellular Mechanisms: Chelation

As a first step against toxic metal exposure, extracellular mechanisms are used to reduce the mobile fraction of metals (Pócsi, 2011). Fungi release chelating compounds like glutathione, siderophores, and organic acids to immobilize metals through the formation of metal complexes or metal salts (Gadd, 2007). The ability of *S. commune* to secrete siderophores and a variety of organic acids could be supported by genomics studies (Kirtzel et al., 2017). Both complexing agents are known to form relatively insoluble metal complexes and salts, thereby immobilizing the respective metal ions and preventing their uptake into the fungal cell (Gadd, 2007). Besides their puta-tive role in metal detoxification, siderophores are also thought to solubilize unavailable forms of heavy metal-bearing minerals by complexation and

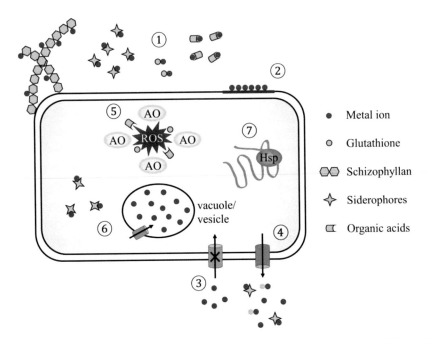

Figure 7.2 Proposed scheme of metal detoxification mechanisms in *S. commune*. Extracellular mechanisms involve chelation of metal ions using siderophores, organic acids such as oxalic and citric acid, schizophyllan, and glutathione (1), and biosorption of metal ions on chitin and pigments (2). Changed metal influx through inhibition of Ca^{2+} and K^+ channels (3) or efflux lead to decreased toxicity (4). Intracellular mechanisms include neutralization of reactive oxygen species (ROS, 5) by overproduction of antioxidants (AO) like glutathione, glutaredoxin, glutathione transferase, and ascorbic acid. Intracellular metal chelation or transport by siderophores or transport of metal-peptide complexes into the vacuole (6), and reduction of metal-induced damage by heat-shock proteins (Hsp, 7).

enhance the metal uptake. A potential role of the siderophores produced by *S. commune* could thus potentially involve both metal mobilization from black slate and metal immobilization.

Organic acids like oxalic, malic, and succinic as well as citric acid are considered to chelate metal ions and could reduce their toxicity in wood-rotting fungi (Gube, 2016). Since total digestion and sequential extraction revealed the existence and bioavailability of metals released from black slate (Wengel et al., 2006), their chelation by organic acids might protect *S. commune* from metal toxicity. However, chelation of metal cations could also change the solution equilibrium between a mineral and its surrounding, which, hence, could cause a further mobilization of ions (Gube, 2016).

7.3.2 Extracellular Mechanisms: Biomineralization

As another detoxification mechanism, oxalic acid could improve the metal tolerance of fungi by precipitating toxic metals as highly insoluble oxalates when interacting with metals or metal-bearing minerals, such as Cd and Pb (Gadd et al., 2012). For *S. commune*, we could demonstrate the formation of the oxalate whewellite during growth with black slate. However, the sole deposition of calcium oxalate and an absence of toxic metal oxalates indicated that metal salts did not play a role in the detoxification mechanism of *S. commune*.

7.3.3 Extracellular Mechanisms: Biosorption

During growth in liquid medium, *S. commune* produced the extracellular polysaccharide schizophyllan. ECMM has excellent toxic metal binding capabilities (Pócsi, 2011), buffers metal diffusion, and promotes metal resistance (Vesentini et al., 2006). Another extracellular mechanism to reduce the toxicity of metals involves their sorption to the cell wall, a passive process called biosorption. The fungal cell wall is mainly composed of chitin, glucan, mannans, and glycoproteins (Freihorst et al., 2016), thus providing a large number of potential metal binding sites like free carboxyl, amino, hydroxyl, mercapto, and other functional groups (Gube, 2016).

During growth on highly concentrated metal plates, *S. commune* produced differently colored pigments. Metal-induced pigment production has also been shown for other fungi (e.g., Baldrian, 2003) that may protect cells from metal toxicity (Gadd, 2007). Especially melanin (brown and black) and carotenoids (yellow, orange, and red) protect fungi from metals by scavenging toxic singlet oxygen (Griffin, 1994). Furthermore, some studies demonstrated a higher biosorption of metals like Cd, Cu, and U in melanized mycelium and assumed that pigments might bind metal ions and immobilize them. Although it is generally known that fungal cells possess high metal sorption capacities, an increasing metal supply could lead to a saturation of hyphal binding sites (Jentschke and Godbold, 2000), and hence biosorption might play a minor role in highly contaminated sites.

7.3.4 Changed Metal Influx and Efflux

Heavy metals enter cells either by diffusion or through channels and transporters which normally facilitate the uptake of essential metal micronutrients such as Fe, Mn, and Zn, anions like phosphate and sulfate as well as sugars and sugar derivatives (Gube, 2016; Pócsi, 2011). Elimination of these channels or transporters could be an efficient and simple way to protect cells from the uptake of toxic metals, and for *S. cerevisiae*, the uptake of As, Cd, and Cr via such transporter was proven (Pócsi, 2011). However, multiple transporters might exist which channel metal ions into the cytoplasm, and eliminating even one of these transport routes could lead to nutrient starvation (Gube, 2016). Microarray analyses revealed downregulations of Ca^{2+} and K^+ channels, including the Ca^{2+}-modulated non-selective cation channel polycystin (4.1-fold), a Ca^{2+}/calmodulin-dependent protein kinase (5.1-fold), a voltage-gated shaker-like K^+ channel (8-fold), and the K^+-channel ERG (ether-a-go-go-related gene; 4.8-fold) (Kirtzel et al., 2019).

To decrease metal concentrations inside the cell, fungi possess efflux systems which export metal ions or their chelates out of the cell. Multidrug transporter export xenobiotics and metals from cells and are responsible for multidrug resistance (Moriyama et al., 2008). Black slate induced the upregulation of transporter genes (14-fold) and proteins (Kirtzel et al., 2017, 2018).

7.3.5 Intracellular Protection Mechanisms

Intracellular siderophores of filamentous fungi were shown to chelate metals and keep excess iron and, albeit with lower binding abilities, other metals such as Cd and Pb in a thermodynamically inert state, hence reducing harmful consequences of free iron ions in the cytosol or cell organelles (Pócsi, 2011). Intracellular chelation and sequestration of metals can also be achieved by phytochelatins, which are synthesized from glutathione and metallothioneins. The latter form complexes with metals, especially Cu, transport metal ions, play a role in metal homeostasis, and could function as antioxidants (Hall, 2002). Complexes of phytochelatins with, e.g., Cd, Cu, and Pb were shown to be sequestered into plant and yeast vacuoles with the help of transporters (Cobbett and Goldsbrough, 2002). Thus, phytochelatins and metallothioneins might play crucial roles in the detoxification of and tolerance to (heavy) metal stress; however, they are not produced by white-rot fungi (Baldrian, 2003). Indeed, no expression was found for both peptides and proteins in the microarray, but a black slate induced regulation of several ABC transporters indicated an increased transport of toxic metals (Kirtzel et al., 2017, 2018).

Besides an accumulation of Fe in the hyphae, a sequestration of Zn in small granules, probably Zn-containing vesicles (zincosomes) was proven (Kirtzel et al., 2019). Furthermore, Cd and Pb were shown to be sequestered in vacuoles. Both observations are in accordance with other studies showing a sequestration of toxic metals in these cells (Ezaki and Nakakihara, 2012).

Another intracellular mechanism to cope with metal stress involves the repair of stress-damaged proteins by metallothioneins and heat shock proteins (HSPs) (Hall, 2002). Proteome and microarray data revealed an upregulation of several HSPs including DnaJ (7-fold), Hsp20 (6.7-fold), and HSS1 of Hsp70 family (2.1-fold) (Kirtzel et al., 2017, 2018).

7.4 Conclusions

Fungi, especially laccase producing white-rot basidiomycetes like *S. commune*, can significantly contribute to the global biogeochemical cycling of elements and, especially when colonizing former mining areas, can cause environmental problems by releasing toxic concentrations of metals. *S. commune* interacting with low-grade metamorphic black slate derived from a former alum mine confirmed the ability of the fungus to degrade this rock using several bioweathering mechanisms. By means of indirect and direct biomechanical as well as several biochemical degradation strategies, *S. commune* was shown to possess a great potential to decay rocks. The presence of fungal-induced etchings and the formation of the biomineral whewellite suggested a close fungus–rock interaction and emphasized the effectiveness of especially biochemical weathering processes.

Examination of biomolecular mechanisms demonstrated a secretion and upregulation of many wood-rotting enzymes which are capable to degrade and utilize organic matter from black slate. These results provide further understanding of the potential roles of basidiomycetes in the biodeterioration of rocks rich in organic matter. A short-term stimulation of fungal metabolism and biomass production due to released nutrients could be shown, while the simultaneously released toxic metals finally restricted fungal growth.

At the same time, metal stress protection, cell wall biosorption, biomineralization, and intracellular protective mechanisms including production of siderophores in concert with changed metal influx and efflux balance the effective metal load. This process-oriented knowledge could be used to improve bioremediation strategies using fungal mycelium to chemically modify metals, remove them by biosorption, and improve the degradation of pollutants by the diverse enzyme machinery of white-rot fungi.

Acknowledgements

We would like to thank the DFG for support through GRK 1257 and JSMC.

References

Ahmed, E., Holmström, S.J.M., 2014. Siderophores in environmental research: Roles and applications. Microb. Biotechnol. 7, 196–208. https://doi.org/10.1111/1751-7915.12117

Baldrian, P., 2003. Interactions of heavy metals with white-rot fungi. Enzyme Microb. Technol. 32, 78–91. https://doi.org/10.1016/S0141-0229(02)00245-4

Burford, E.P., Kierans, M., Gadd, G.M., 2003. Geomycology: Fungi in mineral substrata. Mycologist 17, 98–107. https://doi.org/10.1017/S0269-915X(03)00311-2

Burgstaller, W., Schinner, F., 1993. Leaching of metals with fungi. J. Biotechnol. 27, 91–116. https://doi.org/10.1016/0168-1656(93)90101-R

Cobbett, C., Goldsbrough, P., 2002. Phytochelatins and metallothioneins: Roles in heavy metal detoxification and homeostasis. Annu. Rev. Plant Biol. 53, 159–182. https://doi.org/10.1146/annurev.arplant.53.100301.135154

Ehrlich, H.L., 1998. Geomicrobiology: its significance for geology. Earth Sci. Rev. 45, 45–60. https://doi.org/10.1016/S0012-8252(98)00034-8

Erdmann, S., Freihorst, D., Raudaskoski, M., Schmidt-Heck, W., Jung, E.M., Senftleben, D., Kothe, E., 2012. Transcriptome and functional analysis of mating in the basidiomycete *Schizophyllum commune*. Eukaryot. Cell 11, 571–589. https://doi.org/10.1128/EC.05214-11

Ezaki, B., Nakakihara, E., 2012. Possible involvement of GDI1 protein, a GDP dissociation inhibitor related to vesicle transport, in an amelioration of zinc toxicity in *Saccharomyces cerevisiae*. Yeast 29, 17–24. https://doi.org/10.1002/yea.1913

Fischer, C., Schmidt, C., Bauer, A., Gaupp, R., Heide, K., 2009. Mineralogical and geochemical alteration of low-grade metamorphic black slates due to oxidative weathering. Chem. Erde Geochem. 69, 127–142. https://doi.org/10.1016/j.chemer.2009.02.002

Fomina, M.A., Alexander, I.J., Colpaert, J.V., Gadd, G.M., 2005. Solubilization of toxic metal minerals and metal tolerance of mycorrhizal fungi. Soil Biol. Biochem. 37, 851–866. https://doi.org/10.1016/j.soilbio.2004.10.013

Freihorst, D., Fowler, T.J., Bartholomew, K., Raudaskoski, M., Horton, J.S., Kothe, E., 2016. 13 The mating-type genes of the basidiomycetes, in: Wendland, J. (Ed.), Growth, Differentiation and Sexuality, The Mycota. Springer International Publishing, Cham, pp. 329–349. https://doi.org/10.1007/978-3-319-25844-7_13

Gadd, G.M., 1999. Fungal production of citric and oxalic acid: Importance in metal speciation, physiology and biogeochemical processes. Adv. Microbial Physiol. 41, 47–92. https://doi.org/10.1016/S0065-2911(08)60165-4

Gadd, G.M., 2007. Geomycology: Biogeochemical transformations of rocks, minerals, metals and radionuclides by fungi, bioweathering and bioremediation. Mycol. Res. 111, 3–49. https://doi.org/10.1016/j.mycres.2006.12.001

Gadd, G.M., Rhee, Y.J., Stephenson, K., Wei, Z., 2012. Geomycology: Metals, actinides and biominerals. Environ. Microbiol. Rep. 4, 270–296. https://doi.org/10.1111/j.1758-2229.2011.00283.x

Griffin, D.H., 1994. Fungal Physiology. Wiley, New York.

Gube, M., 2016. 4 Fungal molecular response to heavy metal stress, in: Hoffmeister, D. (Ed.), Biochemistry and Molecular Biology, The Mycota. Springer International Publishing, Cham, pp. 47–68. https://doi.org/10.1007/978-3-319-27790-5_4

Hall, J.L., 2002. Cellular mechanisms for heavy metal detoxification and tolerance. J. Exp. Bot. 53, 1–11. https://doi.org/10.1093/jexbot/53.366.1

Hofrichter, M., Ziegenhagen, D., Sorge, S., Ullrich, R., Bublitz, F., Fritsche, W., 1999. Degradation of lignite (low-rank coal) by ligninolytic basidiomycetes and their manganese peroxidase system. Appl. Microbiol. Biotechnol. 52, 78–84. https://doi.org/10.1007/s002530051490

Huang, W.H., Keller, W.D., 1972. Organic acids as agents of chemical weathering of silicate minerals. Nat. Phys. Sci. 239, 149–151. https://doi.org/10.1038/physci239149a0

Jaffe, L.A., Peucker-Ehrenbrink, B., Petsch, S.T., 2002. Mobility of rhenium, platinum group elements and organic carbon during black shale weathering. Earth Planet. Sci. Lett. 198, 339–353. https://doi.org/10.1016/S0012-821X(02)00526-5

Javadi Nooshabadi, A., Hanumantha Rao, K., 2014. Formation of hydrogen peroxide by sulphide minerals. Hydrometallurgy 141, 82–88. https://doi.org/10.1016/j.hydromet.2013.10.011

Jentschke, G., Godbold, D.L., 2000. Metal toxicity and ectomycorrhizas. Physiol. Plant. 109, 107–116. https://doi.org/10.1034/j.1399-3054.2000.100201.x

Kirtzel, J., Madhavan, S., Wielsch, N., Blinne, A., Hupfer, Y., Linde, J., Krause, K., Svatoš, A., Kothe, E., 2018. Enzymatic bioweathering and metal mobilization from black slate by the basidiomycete *Schizophyllum commune*. Front. Microbiol. 9. https://doi.org/10.3389/fmicb.2018.02545

Kirtzel, J., Scherwietes, E.L., Merten, D., Krause, K., Kothe, E., 2019. Metal release and sequestration from black slate mediated by a laccase of *Schizophyllum commune*. Environ. Sci. Pollut. Res. 26, 5–13. https://doi.org/10.1007/s11356-018-2568-z

Kirtzel, J., Siegel, D., Krause, K., Kothe, E., 2017. Stone-eating fungi: Mechanisms in bioweathering and the potential role of laccases in black slate degradation with the basidiomycete *Schizophyllum commune*. Adv. Appl. Microbiol. 99, 83–101. https://doi.org/10.1016/bs.aambs.2017.01.002

Lee, M.R., Parsons, I., 1999. Biomechanical and biochemical weathering of lichen-encrusted granite: Textural controls on organic–mineral interactions and deposition of silica-rich layers. Chem. Geol. 161, 385–397. https://doi.org/10.1016/S0009-2541(99)00117-5

Leonowicz, A., Cho, N., Luterek, J., Wilkolazka, A., Wojtas-Wasilewska, M., Matuszewska, A., Hofrichter, M., Wesenberg, D., Rogalski, J., 2001. Fungal laccase: properties and activity on lignin. J. Basic Microbiol. 41, 185–227. https://doi.org/10.1002/1521-4028(200107)41:3/4<185::AID-JOBM185>3.0.CO;2-T

Martínez, A.T., Speranza, M., Ruiz-Dueñas, F.J., Ferreira, P., Camarero, S., Guillén, F., Martínez, M.J., Gutiérrez, A., del Río, J.C., 2005. Biodegradation of lignocellulosics: microbial, chemical, and enzymatic aspects of the fungal attack of lignin. Int. Microbiol. Off. J. Span. Soc. Microbiol. 8, 195–204.

Money, N.P., Davis, C.M., Ravishankar, J.P., 2004. Biomechanical evidence for convergent evolution of the invasive growth process among fungi and oomycete water molds. Fungal Genet. Biol. 41, 872–876. https://doi.org/10.1016/j.fgb.2004.06.001

Ohm, R.A., de Jong, J.F., Lugones, L.G., Aerts, A., Kothe, E., Stajich, J.E., de Vries, R.P., Record, E., Levasseur, A., Baker, S.E., Bartholomew, K.A., Coutinho, P.M., Erdmann, S., Fowler, T.J., Gathman, A.C., Lombard, V., Henrissat, B., Knabe, N., Kües, U., Lilly, W.W., Lindquist, E., Lucas, S., Magnuson, J.K., Piumi, F., Raudaskoski, M., Salamov, A., Schmutz, J., Schwarze, F.W.M.R., vanKuyk, P.A., Horton, J.S., Grigoriev, I.V., Wösten, H.A.B., 2010. Formation of mushrooms and lignocellulose degradation encoded in the genome sequence of *Schizophyllum commune*. Nature Biotech. 28, 957–963.

Papida, S., Murphy, W., May, E., 2000. Enhancement of physical weathering of building stones by microbial populations. Int. Biodeterior. Biodegrad. 46, 305–317. https://doi.org/10.1016/S0964-8305(00)00102-5

Pócsi, I., 2011. Toxic metal/metalloid tolerance in fungi—A biotechnology-oriented approach, in: Banfalvi, G. (Ed.), Cellular Effects of Heavy Metals. Springer Netherlands, Dordrecht, pp. 31–58. https://doi.org/10.1007/978-94-007-0428-2_2

Raper, J.R., 1966. Genetics of sexuality in higher fungi. Ronald Press Co., New York.

Schröder, I., Johnson, E., de Vries, S., 2003. Microbial ferric iron reductases. FEMS Microbiol. Rev. 27, 427–447. https://doi.org/10.1016/S0168-6445(03)00043-3

Seifert, A.-G., Trumbore, S., Xu, X., Zhang, D., Kothe, E., Gleixner, G., 2011. Variable effects of labile carbon on the carbon use of different microbial groups in black slate degradation. Geochim. Cosmochim. Acta 75, 2557–2570. https://doi.org/10.1016/j.gca.2011.02.037

Vesentini, D., Dickinson, D.J., Murphy, R.J., 2006. Analysis of the hyphal load during early stages of wood decay by basidiomycetes in the presence of the wood preservative fungicides $CuSO_4$ and cyproconazole. Holzforschung 60, 637–642.

Warscheid, T., Krumbein, W.E., 1994. Mikrobielle Werkstoffzerstörung – Simulation, Schadensfälle und Gegenmaßnahmen für anorganische nichtmetallische Werkstoffe: Biodeteriorationsprozesse an anorganischen Werkstoffen und mögliche Gegenmaßnahmen. Mater. Corros. 45, 105–113. https://doi.org/10.1002/maco.19940450207

Wengel, M., Kothe, E., Schmidt, C.M., Heide, K., Gleixner, G., 2006. Degradation of organic matter from black shales and charcoal by the wood-rotting fungus *Schizophyllum commune* and release of DOC and heavy metals in the aqueous phase. Sci. Total Environ. 367, 383–393. https://doi.org/10.1016/j.scitotenv.2005.12.012

Wösten, H.A.B., 2001. Hydrophobins: Multipurpose proteins. Annu. Rev. Microbiol. 55, 625–646. https://doi.org/10.1146/annurev.micro.55.1.625

8

Biotechnology for the Recovery of Metals using Agroindustrial Wastes

Luz Irene Rojas Avelizapa[1,*], Ricardo Serna Lagunes[1]
and Norma Gabriela Rojas Avelizapa[2]

[1]Facultad de Ciencias Biológicas y Agropecuarias-Córdoba, Universidad
Veracruzana, Josefa Ortiz de Domínguez s/n. Peñuela, 94945, Amatlán de los
Reyes, Veracruz
[2]Centro de Investigación en Ciencia Aplicada y Tecnología Avanzada del
Instituto Politécnico Nacional, Colinas del Cimatario, 76090, Querétaro,
Querétaro, México
E-mail: luzrojas@uv.mx
*Corresponding Author

8.1 Introduction

Over time, industries have discharged toxic compounds into water bodies,
air, and soil, which contain some metals such as copper (Cu), zinc (Zn), lead
(Pb), mercury (Hg), and arsenic (As). These accumulate in organisms and
can easily enter through the skin, the gastrointestinal and respiratory tracts,
causing serious health problems, decreased production of red and white blood
cells, skin changes, and irritation of the lungs, infertility and abortions in
women, cancer of the skin, lungs, liver, as well as central nervous system
conditions and accumulate in the brain (Montero-Álvarez, 2010).

8.2 Agroindustrial Wastes of Interest for the Absorption of Metals

Biotechnology is defined as a technology where organisms, biological systems, or their derivatives are applied for generating or modifying products
or processes for specific uses. It is also considered as a multidiscipline,

in which the cell biology, molecular, microbiology, and analysis tools of the bioinformatics area are applied to carry out research, development of bioactive substances, and functional foods; although, recently this field has expanded to biosecurity, as it is the application of biotechnology to regulate, balance, and restore the environment (Plein, 1991).

Given that biotechnology covers different perspectives and application forms, bioabsorption is a process that currently is evolving by using a method based on biomass *in vivo* (microbial flora, algae, plants, residual biomass, agroindustrial product, and biopolymers) have the objective of capturing, recovering, and/or removing polluting metals such as chromium, nickel, cadmium, lead, and mercury to mention a few. On the other hand, there are a great variety of materials, which have negative effects on the environment and therefore with human health, and those were the ones that excelled for being studied (Tejada-Tovar et al., 2015a).

8.3 Types of Agroindustrial Wastes Used in the Absorption of Metals

In the biotechnology field, the absorption of metals has increased in recent years, which has led to test different types of agroindustrial and biological wastes in order to identify the most suitable biomasses for one or several metals.

As an example, there have been experiments with orange peels (*Citrus sinensis*) as absorbent of Cr (III) and Cr (VI) which are highly toxic, mutagenic, and carcinogenic; it has been found that this biomass can absorb up to 66.6% of Cr (VI) in 120 min, and therefore it is considered that the orange peel can treat wastewater from tanneries and other industries (Tejada-Tovar et al., 2015b). The orange skin, as an agroindustrial waste, has a high potential as a bioabsorbent of Pb (II) and Zn (II) with a 99.5% removal (Cardona-Gutiérrez et al., 2013). It has eliminated up to 74.5% of Lanasol Navy CE commercial colorant (Vargas-Rodríguez et al., 2009), whereas when orange peel is mixed with red algae and prickly pear, Cd, Pb, and Zn are removed (Mendoza y Molina, 2015). Phytoremediation is a way to extract metals using different plant species. For example, maize and Stevia compost have been used to recover degraded soils from heavy metal contamination. Corn absorbs metals from the soil as lead and cadmium in its root, considering it a stabilizing plant for these metals, while the vermicompost of Stevia was more effective in the absorption of heavy metals from the soil (Munive et al., 2018). Other experiments have compared corn husk and orange peel to determine the competitive absorption on Ni (II) and Pb (II) in binary solution, finding that in the first one absorbs in higher concentration with corn husk and the second one with husk of orange (Tejada 2015b). In another study, the palm bagasse with citric acid allows the elimination of Pb (II) in 10 min (Tovar et al., 2015a; Figure 8.1).

Phytoremediation has been contextualized as a tool to eliminate contamination by heavy metals in aqueous environments mainly. With *Pistia stratiotes* commonly known as water lettuce, Cu (III) can be bioabsorbed eliminating it up to 70% in only 6 h (Torres et al., 2007). By using water hyacinth (*Eichhornia crassipes*), it was possible to remove 94.68% of mercury in a wetland for 3 h, but in a continuous artificial surface wetland, up to 99.5% of mercury was removed (Paredes and Ñique, 2015). On the other hand, fungi – *Penicillium* spp. – have been used to remove lead, cadmium, and mercury in mining effluents since they are considered to be dangerous and toxic heavy metals (Sánchez et al., 2014).

The use of microbial biomass (yeasts, fungi, algae, and bacteria) for the detoxification and recovery of toxic or valuable metals present in industrial wastewater, it is based on bioprecipitation, a strategy to concentrate metals and accumulate them within the microbial structure (Cañizares-Villanueva, 2000).

Activated carbon has been obtained through agroindustrial wastes such as corn and sugarcane bagasse, corn husk, coconut, palm, chickpea, pistachio, peanut and walnut, olive and cherry stones, rice bran oil, wasted jaca shell,

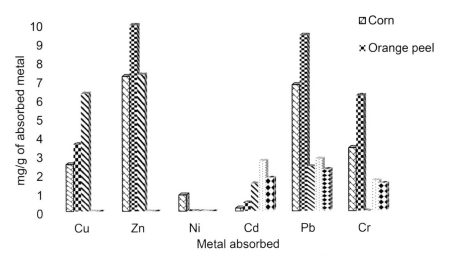

Figure 8.1 Biomass commonly used in the sorption of heavy metals.

rubber seed, cotton stem, fruits such as mombin plum and tea waste, and these have been determined to be excellent precursors because of their high carbon content and low ash content (Mejía, 2018). In this sense, activated charcoal from African palm (*Elaeis guineensis*) has been used to reduce the concentration of Fe^{+3} in drinking water from wells for human use, concluding that the endocarp of African palm as a raw material is suitable for use in obtaining a specific activated carbon for the treatment of industrial waters since it can remove up to 0.21 mg/L of Fe^{+3} (Ávalos et al., 2018).

It has been assessed that sorghum is a good bioabsorbent that has been tested in the treatment of agricultural waste dyes, which are anionic in nature, since it has been determined that it is capable of extracting a number of functional and superficial carbon groups and oxygen (Galindo et al., 2018). While the pineapple skin has been used as an absorbent of red 40, a typical dye in the food industry, removing up to 84% of this disodium salt (Durango et al., 2018), which is presented as an additive in sweets, dairy products, cookies, gelatin, condiments, drinks, desserts, cake mixes, and fruit flavored fillings, that are associated with numerous forms of cancer, hyperactivity, and behavioral problems in children.

8.4 Mechanisms Involved in the Removal of Metals

Experiments have been conducted with different levels of treatment with almost all sources of variation (types of waste) and probabilities at different levels of exclusion (temperature gradients, different particle sizes, time/time of operation, and pH) for determining the best treatment which absorbs a greater amount of metals in less time. These mechanisms in the absorption process have been modified based on the need to remove heavy metals; therefore, there is no standard technique for bioresorbable metals of industrial effluents as seen in mining, with electroplating or precious metals by solutions in industrial processes. Thus, the biosorption mechanisms are standardized after testing a wide range of conditions until the optimal absorption mechanism is determined (Sala et al., 2010).

Among the microorganisms involved in the process of absorption of heavy metals, mention may be made about *Penicillium*, *Aspergillus*, and *Paecilomyces* as living organisms, while as biomass, it has been used in tamarind, orange, apple, and barley, to mention a few; as biopolymers, bentonite-chitosan and chitosan-epichlorohydrin triphosphate have been used mainly. In another context, activated charcoal derived from *E. coli*, from *Arthrobacter viscous*, from orange peel and coconut has been used, with some chemical

Table 8.1 Bioabsorption studies using different biomass and parameters used in the experimental design

Biomass	Particle Size (μm)	Temperature (°C)	pH	Metal/ Substance	Author(s)
Manihot esculenta	0,297; 0,149 y 0,105	30, 40 y 50	–		Brousse et al. (2012)
Citrus cinensis	0.6, 0.42 y 0.3	25	2, 3, 4, 5, and 6	Cr (III)	Pinzón-Bedolla et al. (2010)
Rice husk	0.25–0.75	25	5–9	Azul de metileno	Moreno et al. (2013)
Coffea arabica	0.5	10–20	3.5–5	Pb (II) y Cd (II)	Pacheco-Tanaka et al. (2010)
Sheet of *Coffea arabica*	0.14, 0.21 y 1.68	25	1–5	Cr (IV)	Cobos et al. (2009)
Olive and ramón olive tree	0.5 a 1.0	25	3.11	Pb (II)	De Castro et al. (2008)

modifications such as the addition of glutaraldehyde, calcium chloride, or citric acid (Tejada-Tovar et al., 2015b).

The kinetics of metal absorption has focused on determining what the particle size is, temperature and pH suitable to absorb different metals or elements. Different studies show the different measures of particle size, temperature, and pH that apparently vary depending on the metal to be absorbed (Table 8.1). In different studies, it has been reported that pH is a determinant to increase the absorption capacity; it means that there is an inverse relationship between pH and absorption capacity. For example, in corn crops when the pH is greater than 7, the absorption capacity of Cd and Zn is increased by 20%; this is because at this pH, the solubility of the metals is controlled by carbonates and minerals present in soil (Carrillo-González et al., 2003), while in barley cultures (*Hordeum vulgare*) the absorption of Cr^{+6} occurs in conditions of pH 6–7.5, the latter being the optimum with which a greater amount of Cr^{+6} is absorbed (Ramón-Jara, 2017), but in an aqueous medium, the Pb (II) of wastewater is absorbed in a higher concentration at pH 5 using corncob (Oré-Jiménez et al., 2015). When comparing the effectiveness of native chitosan particles in sizes lesser than 75 μm and greater than 2000 μm, it was determined that particles smaller than 75 μm have the same absorption capacity of Cu (II) compared to beads (qm $\sim$200 mg G^{-1}; Flores et al., 2005).

On the other hand, the particle size represents an important measure for the bioabsorption process, and because of identifying the particle size and the appropriate pH, the costs of the process can be reduced and the efficiency in the absorption times is increased (Jara-Peña et al., 2014). For example, the optimum particle size of maize zuro (0.5 mm) and orange peel (1 mm) is obtained up to 67.5% and 99.2% of Pb (II) removal (Tovar et al., 2016).

In several studies, the removal of heavy metals using agroindustrial residues is shown and this concentration can be variable depending on the experimental conditions and the levels or sources of variation in the metal removal experiments, but the amount of heavy metal absorption depends on the kinetics of absorption and the bioabsorption capacity of biomass (Cuevas and Walter, 2004). In Figure 8.1, the different biomasses used for the absorption of different heavy metals are shown. Among the advantages of using agroindustrial wastes in the removal and/or absorption of heavy metals is the high availability of biomass of an agroindustrial nature that are available, the varied chemical composition of the biomass that gives it a high adsorption capacity, and, in general, the low cost of this biomass that can be reused in several cycles of absorption, where the metals are attached to the cell wall surface and accumulate inside the cells, but the most important is that eliminating organometallic compounds can be done economically (He and Chen, 2014).

8.5 Biopolymers as Effective Adsorbents in the Removal of Metals

Heavy metal contamination is a serious problem due to its toxicity even at low concentrations. Heavy metals are not biodegradable and studies have shown that they accumulate in the body because they are not discarded naturally, which results in health alterations of all living beings that are exposed to these metals (Crini, 2005). For this reason, the convenience of removing or reducing the concentrations of these heavy metals has been noted, especially in aqueous media, for which different methods and technologies have been used. Generally, the methods include precipitation, ion exchange, adsorption, and processes through membranes. However, some of these methods have the disadvantage of generating by-products which may be more hazardous or that require further treatment (Gavrilescu, 2004; Reddad et al., 2002).

One of the processes that has shown the best results in the removal of metals is the adsorption; however, when using adsorbent materials such as activated carbon or ion exchange resins, mainly in the treatment of wastewater, these represent a very high cost.

The commercial polymers and ion exchange resins used in decontamination processes come from materials made from petroleum, through chemical

processes that are harmful to the environment. The use of natural polysaccharides as adsorbents, which have proved to be a good alternative in the removal
of metal ions, has also been proposed. The main characteristic of these natural
adsorbent materials is their high availability in nature or as an industrial waste
that can be recycled and exploited for this purpose (Bailey et al., 1999).

The biopolymers are formed by monosaccharides and are characterized
by being biodegradable, biocompatible, polyfunctional, hydrophilic, and non-
toxic and have a high chemical reactivity. Also, it is known that they have the
ability to associate a wide variety of molecules through physical and chemical
interactions (Crini, 2005).

8.5.1 The Chitosan Biopolymer

Chitin is a very abundant polysaccharide in nature, mainly in crustaceans,
insects, and fungi. It has a linear structure of high molecular weight consti-
tuted by units of N-acetyl-D-glucosamine linked by β-D bonds (1,4). It is
highly insoluble and has low reactivity. Partial deacetylation of chitin gives
rise to chitosan, which exhibits better reactivity and solubility.

Chitosan can then be defined as a linear polysaccharide composed of
randomly distributed chains of β-(1-4) d-glucosamine (deacetylated units)
and N-acetyl-D-glucosamine (acetylated unit) (Figure 8.2).

One of its many applications is its high affinity for metal ions, especially
heavy metals of the transition series, which is due to the large number of
amino groups located in the two positions of the glycosidic rings; it acts as
primary amine and develops a free and effective action on the pair of unpaired
electrons (Flores et al., 2001).

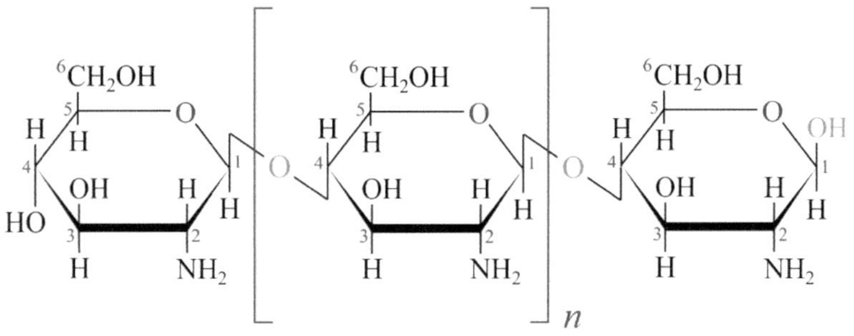

B-(1,4)-D-glucosamine

Figure 8.2 Chemical structure of chitosan.

Chitosan interaction with metal ions is a complex mechanism involving the chelation of these ions in solutions close to neutral, ion exchange, and adsorption. For the chelating effect of chitosan to take place on metal ions, the -OH and -O groups of the D-glucosamine residues are required as binders, and at least 2 or more amino groups of the same chain are required to join the same metallic ion (Cartaya, et al., 2009; Dou, HY et al., 2013).

The properties of chitosan lead to the protonation of amino groups at acid pHs. These cationic properties make the polymer very efficient in the adsorption of metal ions by electrostatic interactions (Montero-Álvarez, 2010).

The interaction ability of chitosan with heavy metals was mentioned for the first time by Chui et al. (1996), arguing that the amino sugars of chitin and chitosan have effective binding sites with metal ions, forming stable coordination complexes (Mora, 2012).

Trimukhe and Varma (2008) used the cross-linked chitosan to form complexes with heavy metals and chitosan with good results in the removal of iron, copper, and lead. Song et al. (2007) using cellulose with chitosan grafts as an exchange resin to remove heavy metals from water, which showed high selectivity at low concentrations. Whereas Aroua et al. (2007) used chitosan in solution to complex metals with 50% maximum removal. Paulino et al. (2007) used chitosan with different degrees of acetylation to absorb lead and nickel, showing specific absorptions. The state-of-the-art of the use of chitosan in the removal of heavy metals is extensive, estimating that it can interact with heavy metals in a specific way and presenting a wide range of formulations (Dou, H. Y. et al., 2013). In this regard, studies have been made; these studies determine that the size of the particle is very important for the capacity of adsorption, which is why it has been physically modified to use it as dust, nanoparticles and gels, the latter as pearls, membranes, sponges, honeycombs, etc.

A small particle size decreases the crystallinity of the material and improves the removal capacity of the adsorbent (Qi and Xu, 2004). Chemical modifications have also been made to improve its adsorption and solubility properties in water or acid, performing substitution reactions, depolymerization (chemical, physical, and enzymatic) and elongation of the polymer chain (cross-linking, grafting, copolymerization, and polymer networks) (Wu et al., 2012).

Cross-linking reactions have been shown to improve mechanical and chemical resistance. Some cross-linking agents that have been used are glutaraldehyde, epichlorohydrin, and ethylene glycol diglycidyl ether (EGDE) for obtaining chitosan beads (Wan Ngah et al., 2004). Table 8.2 presents some relevant studies on the removal of metals by adsorption with chitosan

Table 8.2 Adsorption of metals/ions using modified chitosan adsorbents by different authors during 2012–2015

Modified Chitosan Adsorbent	Metals/ions	Efficiency
Chitosan grafted with 3,4-dimethoxy-benzaldehyde	Cd(II)	217.4 mg/g
Chitosan modified multiwalled carbon nanotubes	U(VI)	71 mg/g
Xanthate-modified magnetic cross-linked chitosan	Co(II)	18.5 mg/g
Ethylene-1,2-diamine-6-deoxy-chitosan	Cu(II), Pb(II), Zn(II)	
Ethylene-1,2-diamine-6-deoxy-N-phthaloylchitosan	Cu(II), Pb(II), Zn(II)	32.3, 28.6, 18.6 mg/g
Magnetically modified graphene oxide-chitosan composite	Cr(VI)	82 mg/g
Magnetic cross-linked chitosan grafted with tetraethylenepentamine	UO2(II)	486 mg/g
Cross-linked chitosan modified with histidine	Ni(II)	55.6 mg/g
Protonated chitosan beads	Fe(III)	7.042 mg/g
Carboxymethyl chitosan beads	Fe(III)	9.346 mg/g
Grafted chitosan beads	Fe(III)	14.286 mg/g
Chitosan grafted with 4,4′-diformyl-α-ω-diphenoxy-ethane	Cu(II), Co(II), Zn(II)	12, 8, 12 mg/g
Chitosan grafted with 4,4′-diformyl-α-ω-diphenoxy-ethane	Hg(II), Pb(II)	56, 50 mg/g
Ethylenediamine-modified yeast biomass coated with magnetic chitosan	Pb(II)	121.26 mg/g
Chitosan-thioglyceraldehyde Schiff's base cross-linked magnetic resin	Hg(II), Cu(II, Zn(II)	98, 76, 52 mg/g
Chloroacetic grafted chitosan	Co(II), Cu(II)	59.1, 175.12 mg/g
Glycine grafted chitosan	Co(II), Cu(II)	82.9, 165.91 mg/g
Chitosan cross-linked with glutaraldehyde	Cu(II), Hg(II)	177.8, 661.5 mg/g
Chitosan cross-linked with epichlorohydrin	Cu(II), Hg(II)	146.1, 681.7 mg/g
Amino terminated hyperbranched dendritic polyamidoamine 3rd generation chitosan beads	Cr(VI)	224.2 mg/g
EDTA-modified chitosan	Co(II)	79.7 mg/g
Chitosan grafted with n-butylacrylate	Cr(VI)	17.15 mg/g
Xanthate carboxymethyl grafted chitosan	Cu(II), Ni(II)	174.2, 128.4 mg/g

(Continued)

Table 8.2   Continued

Modified Chitosan Adsorbent	Metals/ions	Efficiency
Cross-linked chitosan with citric acid	Pb(II)	101.7 mg/g
Montmorillonite modified with chitosan	Co(II)	150 mg/g
Triethylene-tetramine modified magnetic chitosan	Th(IV)	133.3 mg/g
Diethylenetriamine-functionalized magnetic chitosan	U(VI)	65.16 mg/g
Magnetic chitosan grafted with α-ketoglutaric acid	Cd(II)	201.2 mg/g
Magnetic chitosan	Hg(II)	155 mg/g
Chitosan grafted with itaconic acid	Cd(II), Pb(II)	405, 334 mg/g

8.5.2 Advantages of the use of Chitosan as an Adsorbent for the Removal of Metals

Chitosan exhibits excellent selectivity towards metals, dyes, aromatic, and phenolic compounds. The adsorption of heavy metals by chitosan is a low-cost process since the raw material from which the biopolymer is obtained is a waste material that is generated in large quantities. It has excellent adsorption capacities: greater than 1 mmol/g for most metals and for dyes and this capacity is 3–15 times higher compared to commercial activated carbons as shown in Tables 8.1 and 8.2, which summarize the results obtained by different authors (Liu et al., 2013; Kyzas and Bikiaris, 2015).

References

Aroua, M.K., Zuki, F.M., Sulaiman, N.M., 2007. Removal of chromium ions from aqueous solutions by polymer-enhanced ultrafiltration. J. Hazard. Mater. 147, 752–758. https://doi.org/10.1016/j.jhazmat.2007.01.120

Bailey, S.E., Olin, T.J., Bricka, R.M., Adrian, D.D., 1999. A review of potentially low-cost sorbents for heavy metals. Water Res. 33, 2469–2479. https://doi.org/10.1016/S0043-1354(98)00475-8

Barboza, L., Avalos, H., 2018. Efectividad del carbon activado elaborado del endocarpio de palma africana (*Elaeis guineensis*), para disminuir la concentración de Fe^{3+} en el agua potable del pozo tubular de la habilitación urbana municipal de Manantay, Noviembre 2017. Univ. Nac. Ucayali.

Broche-Galindo, M.H., Rodríguez-Rico, I.L., Coca-Rives, Y., Calero de Hoces, M. 2018. Characterization of sorgh agricultural waste for use as coloring material biosorbent. Centro Azúcar 45(4): 64–75.

Brousse, M.M., Nieto, A.B., Linares, A.R., Vergara, M.L., 2012. Cinética de adsorción de agua en purés deshidratados de mandioca (*Manihot esculenta* Crantz): Ingenta Connect Supplementary Data. Rev. Venez. Cienc. Tecnol. Aliment. 3, 80–96.

Cardona Gutiérrez, A.F., Cabañas Vargas, D.D., Zepeda Pedreguera, A., 2013. Evaluación del poder biosorbente de cáscara de naranja para la eliminación de metales pesados, Pb (II) y Zn (II). Ingeniería 17(1), 1–9.

Carrillo-González, R., Cruz-Díaz, J., Cajuste, L.J., 2003. Interacción Zn-Cd en el suelo y maíz. Terra Latinoam. 21, 31–40.

Cartaya, O., Peniche, C., Reynaldo, I., 2009. Polímeros naturales recolectores de iones metálicos. Iberoamericana de polímeros. 10(2), 81–94.

Castro, F.H.B. de García, G.B., Hoces, M.C. de Martín-Lara, M.A., 2008. Influencia de algunas variables en la biosorción de plomo con residuos agrícolas. Afinidad 65(536), 286–292.

Cobos, O.F., Florez García, L.C., Arroyave Londoño, J.F., 2009. Estudio de la biosorción de cromo con hoja de café. Ing. E Investig. 29, 59–64.

Crini, G., 2005. Recent developments in polysaccharide-based materials used as adsorbents in wastewater treatment. Prog. Polym. Sci. 30, 38–70. https://doi.org/10.1016/j.progpolymsci.2004.11.002

Cuevas, G., Walter, I., 2004. Metales pesados en maíz (*Zea mays* L.) cultivado en un suelo enmendado con diferentes dosis de compost de lodo residual. Rev. Int. Contam. Ambient. 20, 59–68.

Dou, H.Y., Chen, X.Q., Li, Z.Q., 2013. A review on the use of chitosan and its derivatives as radical scavenger and metal ions chelating agent. Adv. Mater. Res. 734–737, 2218–2221. https://doi.org/10.4028/www.scientific.net/AMR.734-737.2218

Durango, V.L.U., Vásquez-Noreña, P.A., Zapata, R.B., 2018. Uso de cáscara de piña como adsorbente de rojo 40 (típico de la industria alimentaria). Rev. Colomb. Investig. Agroindustriales 5, 87–95. https://doi.org/10.23850/24220582.1362

Flores, J.A., Ly, M., Tapia, N., Maldonado, H., 2001. Biosorción con quitosano: estudios de equilibrio. Rev. Quím. 15, 133–147.

Flores, J.A., Navarro, A.E., Ramos, K.P., Chang, L., Ale, N., Ly, M., Maldonado, H.J., 2005. Adsorción de Cu (II) por quitosano en polvo y perlas de gel. Rev. Soc. Quím. 71, 17–25.

Gavrilescu, M., 2004. Removal of Heavy metals from the environment by biosorption. Eng. Life Sci. 4, 219–232. https://doi.org/10.1002/elsc.200420026

He, J., Chen, J.P., 2014. A comprehensive review on biosorption of heavy metals by algal biomass: Materials, performances, chemistry, and modeling simulation tools. Bioresour. Technol. 160, 67–78. https://doi.org/10.1016/j.biortech.2014.01.068

Jara-Peña, E., Gómez, J., Montoya, H., Chanco, M., Mariano, M., Cano, N., 2014. Capacidad fitorremediadora de cinco especies altoandi-nas de suelos contaminados con metales pesados. Rev. Peru. Biol. 21(2), 145–154.

Kyzas, G.Z., Bikiaris, D.N., 2015. Recent modifications of chitosan for adsorption applications: A critical and systematic review. Mar. Drugs 13, 312–337. https://doi.org/10.3390/md13010312

Liu, B., Wang, D., Yu, G., Meng, X., 2013. Adsorption of heavy metal ions, dyes and proteins by chitosan composites and derivatives — A review. J. Ocean Univ. China 12, 500–508. https://doi.org/10.1007/s11802-013-2113-0

Mejia, M.V.V., 2018. Potencial de residuos agroindustriales para la síntesis de Carbón Activado: una revisión. Sci. Tech. 23, 411–419.

Mendoza, L., Molina, N., 2015. Biosorción de Cd, Pb y Zn por biomasa pretratada de algas rojas, cáscara de naranja y tuna. Cienc. E Ing. Neogranadina 25, 43–60.

Montero-Álvarez, J.A., Paredes-Bautista, M.J., Rivera-Morales, M.C., 2010. Utilización de quitosana para la remoción de arsénico (As) del agua. Superf. Vacío 23, 136–139.

Mora-Molina, J., Chaves-Barquero, L., Araya-Marchena, M., Starbird-Pérez, R., 2012. Desarrollo de membranas de quitosano y diseño de un equipo para la eliminación de metales pesados del agua. Rev. Tecnol. En Marcha 25. https://doi.org/10.18845/tm.v25i3.453

Moreno, A., Figueroa, D., Hormaza, A., 2013. Adsorción de azul de metileno sobre cascarilla de arroz. Prod. Limpia 7(1), 9–18.

Munive, R., Loli Figueroa, O., Azabache Leyton, A., Gamarra Sánchez, G., 2018. Fitorremediación con Maíz (*Zea mays* L.) y compost de Stevia en suelos degradados por contaminación con metales pesados. Sci. Agropecu. 9, 551–560. https://doi.org/10.17268/sci.agropecu.2018.04.11

Oré-Jiménez, F., Lavado Meza, C., Bendezú Montes, S., 2015. Biosorción de Pb (II) de aguas residuales de mina usando el marlo de maíz (*Zea mays*). Rev. Soc. Quím. Perú 81, 122–134.

Pacheco-Tanaka, M.E., Pimentel Frisancho, J.P., Roque Villanueva, W.F., 2010. Cinética de la bioadsorción de iones cadmio (II) y plomo (II) de soluciones acuosas por biomasa residual de café (*Coffea arabica* L.). Rev. Soc. Quím. Perú 76, 279–292.

Paulino, A.T., Guilherme, M.R., Reis, A.V., Tambourgi, E.B., Nozaki, J., Muniz, E.C., 2007. Capacity of adsorption of Pb^{2+} and Ni^{2+} from aqueous solutions by chitosan produced from silkworm chrysalides in different degrees of deacetylation. J. Hazard. Mater. 147, 139–147. https://doi.org/10.1016/j.jhazmat.2006.12.059

Peredes, S., Ñique, L., 2015. Optimización de la fitorremediación de mercurio en humedales de flujo contínuo empleando *Eichhornia crassipes* "jacinto de agua". Investigación y Amazonía 5(1), 44–49.

Pinzón-Bedoya, M.L., Cardona-Tamayo, A.M., 2010. Influencia del pH en la bioadsorción de Cr(III) sobre cáscara de naranja: Determinación de las condiciones de operación en proceso discontinuo. Bistua Rev. Fac. Cienc. Básicas 8(1), 21–30.

Plein, L.C., 1991. Popularizing biotechnology: The influence of issue definition. Sci. Technol. Hum. Values 16, 474–490.

Qi, L., Xu, Z., 2004. Lead sorption from aqueous solutions on chitosan nanoparticles. Colloids Surf. Physicochem. Eng. Asp. 251, 183–190. https://doi.org/10.1016/j.colsurfa.2004.10.010

Ramón-Jara, F., 2017. Capacidad del residuo de la cebada "*hordeum vulgare*" para la absorción de cromo (Cr^{+6}) en aguas contaminadas a nivel del laboratorio. Thesis. Universidad César Vallejo.

Reddad, Z., Gerente, C., Andres, Y., Le Cloirec, P., 2002. Adsorption of several metal ions onto a low-cost biosorbent: Kinetic and equilibrium studies. Environ. Sci. Technol. 36, 2067–2073. https://doi.org/10.1021/es0102989

Rodríguez, M.V., Vargas, D.C., Marrufo, M.G., Benetton, X.D., 2009. Evaluación del proceso de biosorción con cáscaras de naranja para la eliminación del colorante comercial Lanasol Navy CE en aguas residuales de la industria textil. Ingeniería 13, 39–43.

Sala, L.F., García, S.I., González, J.C., Frascaroli, M.I., Bellú, S., Mangiameli, F., Blanes, P., Mogetta, M.H., Andreu, V., Atria, A.M., Peregrin, J.M.S., 2010. Biosorción para la eliminación de metales pensados en aguas de desecho. An. Quím. 106(2), 114–120.

Sánchez, J., Negrete, J.L.M., Urango, I., 2016. Biosorción simultanea de plomo y cadmio en solución acuosa por biomasa de hongos *Penicillium sp.* Temas Agrar. 19, 63–72. https://doi.org/10.21897/rta.v19i1.725

Song, S.-H., Yeom, B.-Y., Shim, W.S., Hudson, S.M., Hwang, T.-S., 2007. Synthesis of biocompatible CS-g-CMS ion exchangers and their adsorption behavior for heavy metal ions. J. Ind. Engineer. Chem. 13(6), 1009–1016.

Tejada, C., Herrera, A.P., Núñez, J.R., 2015. Adsorción competitiva de Ni (II) y Pb (II) sobre materiales residuales lignocelulósicos. Rev. Investig. Andina 17, 1355–1367.

Tejada Tovar, C., Herrera, A., Núñez Zarur, J., 2016. Remoción de plomo por biomasas residuales de cáscara de naranja (*Citrus sinensis*) y zuro de maíz (*Zea mays*). Rev. UDCA Actual. Divulg. Científica 19, 169–178.

Tejada-Tovar, C., Villabona-Ortiz, Á., Garcés-Jaraba, L., 2015. Adsorción de metales pesados en aguas residuales usando materiales de origen biológico. TecnoLógicas 18, 109–123.

Torres, G., Navarro, A.E., Languasco, J., Campos, K., Cuizano, N.A., 2007. Estudio preliminar de la fitoremediación de cobre divalente mediante *Pistia stratioides* (lechuga de agua). Rev. Latinoam. Recur. Nat. 3, 13–20.

Tovar, C.T., Bolaños, E.Q., Benitez, L.T., Bolivar, W.M., 2015a. Absorción de Cromo Hexavalente en soluciones acuosas por cascaras de naranja (*Citrus sinensis*). Prod. Limpia 10(1), 9–21.

Tovar, C.T., Paternina, E.R., Mercado, J.G., Bohorquez, J.M., 2015b. Evaluación de la biosorción con bagazo de palma africana para la eliminación de Pb (II) en solución. Rev. Prospect. 13, 59–67. https://doi.org/10.15665/rp.v13i1.360

Trimukhe, K.D., Varma, A.J., 2008. Complexation of heavy metals by crosslinked chitin and its deacetylated derivatives. Carbohydr. Polym. 71, 66–73.

Villanueva, R.O.C., 2000. Biosorción de metales pesados mediante el uso de biomasa microbiana. Rev. Latinoam. Microbiol. 42, 131–143.

Wan-Ngah, W.S., Kamari, A., Koay, Y.J., 2004. Equilibrium and kinetics studies of adsorption of copper (II) on chitosan and chitosan/PVA beads. Int. J. Biol. Macromol. 34, 155–161. https://doi.org/10.1016/j.ijbiomac.2004.03.001

Wu, F.-C., Tseng, R.-L., Juang, R.-S., 2002. Adsorption of dyes and humic acid from water using chitosan-encapsulated activated carbon: Adsorption of dyes and humic acid from water. J. Chem. Technol. Biotechnol. 77, 1269–1279. https://doi.org/10.1002/jctb.705

9

Environmental Pollution and Current Bioremediation Strategies for Cadmium Containing Residues

Diana Alexandra Calvo Olvera[1,*] and
Norma Gabriela Rojas Avelizapa[2]

[1]Departamento de Biotecnología, Centro de Investigación en Ciencia Aplicada y Tecnología Avanzada del Instituto Politécnico Nacional, Cerro Blanco 141, Colinas del Cimatario, 76090, Santiago de Querétaro, Querétaro, México
[2]Centro de Investigación en Ciencia Aplicada y Tecnología Avanzada del Instituto Politécnico Nacional, Colinas del Cimatario, 76090, Querétaro, Querétaro, México
E-mail: dcalvoo1500@alumno.ipn.mx
*Corresponding Author

9.1 Introduction

Global contamination by heavy metals has been growing due to the increasing exploitation of natural resources and industrial processes, causing serious damage to the health and ecosystems. Heavy metals are those elements that have metallic properties and a density greater than 5 g/cm^3, with the exception of Ti (Titanium), which has a density of 4.6 g/cm^3 and As (Arsenic), which though it is not considered as metal, it presents a high density (5.7 g/cm^3). These metals cause health and environmental problems at low concentrations due to their high toxicity. The toxicity of heavy metals is related to their mobility in the environment, which depends on their chemical speciation, persistence, and the tendency of accumulation or bioaccumulation, due to this, cadmium (Cd) is one of the most toxic heavy metals along with lead (Pb) and mercury (Hg) (Covarrubias and Cabriales, 2017; Rebollo, 2012).

157

9.2 Environmental Pollution by Cadmium

Cadmium is a chemical element that is always associated with some other elements, usually with zinc, Sphalerite (ZnC); it can be found with other metals such as lead and copper or with elements such as oxygen, chlorine, or known sulfur like Greenockite (CdS). Its symbol is Cd, its atomic number is 48, and its density is 8.65 g/cm^3; it is a bluish-white ductile metal and together with zinc it is part of the Earth's crust, so it is released to the environment naturally through volcanic eruptions, rock decomposition, and hydrothermal vents. However, another large part of it is obtained by anthropogenic activities, such as incineration of urban wastes, manufacture, and application of phosphate fertilizers (because they are made with phosphoric rocks as apatite that reaches up to 500 mg/kg of Cd), mining, metallurgy, industrial activities, nickel–cadmium battery waste and in some pigments (Wright, 2003). It is estimated that every year between 25,000 and 30,000 tons of Cd are released into the environment globally, polluting the air, soil, and bodies of water. Its environmental dispersion depends on factors such as redox potential, pH, amount of organic matter, and the presence of clays and iron oxides.

The greatest source of natural emission of Cd into the atmosphere is volcanic activity, while the emissions from mining, refining, and smelting of minerals are the anthropogenic activities that release the greatest amount of Cd into the atmosphere, which is equivalent to approximately 7300 tons per year (Caviedes et al., 2015; Martelli et al., 2006). Cadmium is one of the most toxic metals because it has four of the main characteristics to be considered toxic: the harmful effects for humans and the environment, bioaccumulation, long life time, and it can travel long distances in the wind and in streams of water.

Cadmium has two main routes of entry to the human body: the first is through the intake of food or water and the second through inhalation. In the case of food, these can absorb Cd present in the soil or water; in the soil, the Cd is absorbed by the organic matter present and, in turn, it is absorbed by the plants and deposited in their tissues. In acidic soils, Cd absorption by plants increases, and when these plants are ingested by animals or humans, it enters the body bioaccumulating into it. Some examples of food that may contain considerable amounts of Cd include mushrooms, cocoa, fish, crustaceans, and the meat of ruminants, mainly the kidney and liver.

Another problem to consider with Cd in the soil is that it alters the soil structure and microbiological processes that occur there; some decomposer animals such as earthworms are sensitive to this metal and die at low concentrations, while at high concentrations, it can reach and kill soil microorganisms, threatening the entire ecosystem. Cd is also accumulated by its intake through water or aquatic organisms consumed by human beings; water contamination by Cd can be caused by wastewater from industries, runoff from agricultural areas, upwelling (transporting dissolved metals from deeper areas to coastal areas), or it can be transported by air until it gets deposited in the water.

In aquatic ecosystems, Cd usually bioaccumulates in filtering organisms such as bivalves, cnidarians, sponges, tunicates, polychaetas, whales and fish or in organisms that feed on them. It is known that saltwater organisms are more resistant to the harmful effects of Cd than freshwater organisms; in both cases, they bioaccumulate Cd in their tissues and reach humans through their consumption. The Official Mexican Standards state that the maximum permissible limit for the consumption of fish and shellfish is a Cd content of 0.5 mg/kg (Frías et al., 2010) and the international agencies in charge of public health care recommend a maximum intake limit daily for an adult of 55 μg of Cd per day, while the maximum permissible rate in water is 1 mg of Cd per liter (Frías et al., 2009).

The second way in which Cd can enter the body is by inhalation; through the smoke of tobacco and air released by landfills of hazardous wastes or factories, Cd is inhaled and taken to the lungs and from there to the blood running through the body until reaching the kidneys where protein binds and accumulates, causing problems in the filtration process and damaging the kidneys. Cadmium can be found in the atmosphere under extreme conditions of oxidation in the form of cadmium oxide (CdO) which is also harmful to health or can be found as chloride or cadmium sulfate, which can travel long distances and be deposited on the soil or bodies of water (Sánchez, 2016).

When the Cd is ingested in food or water, it is usually deposited in the intestine, liver, and kidneys, while when it is inhaled, it accumulates in the upper respiratory tract and lungs; once the Cd is absorbed by the tissues, it has an average life of 20 years in humans (Martelli et al., 2006).

The exposure and bioaccumulation of Cd in mammals causes various health problems; the symptoms presented will depend on the way it enters the body and therefore the target organs. The molecules that trap and carry the Cd to the tissues have not been fully elucidated; however, it is known that metals have a high chemical affinity with proteins, which facilitates their accumulation and hinders degradation and elimination. In addition, Cd uses specific receptors for other essential elements for the body such as calcium, zinc, and iron, using a strategy similar to the Trojan horse to be assimilated by the body. Once in the body, Cd can cause emphysema, anosmia, chronic rhinitis, cancer, anemia, eosinophilia, nephropathy, proteinuria, osteoporosis, and osteomalacia. Several studies have shown that Cd can cross several membranes including the placenta causing iron deficiency in the fetus that leads to a delay in growth and anemia (Carballo et al., 2017; Martelli et al., 2006).

9.3 Cadmium Producing Industries and their Wastes

Cd is used mainly for coating objects to protect them from corrosion; however, with the increase in the use of technology, currently 80% of the Cd produced is destined for the manufacture of rechargeable batteries. Other uses and applications of Cd are: pigment, stabilizers for plastics such as PVC, alloys with copper, silver and aluminum, Cd salts used in photography and lithography, as a hardener of tires for manufacturing photoconductors and nuclear reactors for their capacity to absorb slow neutrons (CAMIMEX, 2012; Frias et al., 2010; Godoy, 2011).

Worldwide, Cd is obtained as a by-product from the smelting and refining of zinc and, to a lesser extent, lead and copper; however, currently the production of secondary or recycled cadmium (mainly nickel–cadmium batteries) accounts for 20% of total production. Among the countries with the highest production are: China with 33%, South Korea 18%, and Japan 8%, while Mexico is in fifth place with 6%, which corresponds to 1600 tons a year, which depend directly on the zinc production. The ratio of zinc to cadmium varies, but they are approximately 0.2%–0.4%, while for copper and lead in proportions of 0.3%–0.5% (CAMIMEX, 2012).

The cadmium production companies with greater relevance are Grupo México and Grupo Peñoles (México), which together export 2.4 million dollars of Cd and its manufactures. According to data from CAMIMEX, from January to July 2016, 658 tons were exported, while in 2017, production was down 8.2% with respect to 2016 due to a drop in the production of the Peñoles group, not due to its competitor which has maintained similar levels in its production, with the Charcas mine in San Luis Potosí of the industrial group Minera México, being the one that contributes the greater production of Cd (CAMIMEX, 2016).

According to data from the United Nations program for the environment in Mexico, between 2001 and 2009, 880,000 tons of Cd products were exported, including fertilizers, fertilizers, phosphates, antioxidants, and stabilizing compounds, and 1.2 million were imported tons (Sánchez and Martínez, 2016). The main sources of Cd contamination are metallurgy, the manufacture of phosphate fertilizers, the incineration of waste, combustion of hydrocarbons, and waste products containing Cd, mainly batteries.

Once Cd is in the soil, it can remain for up to 300 years and it is estimated that approximately 90% of this will be unprocessed, so it is important to reject Cd waste and to implement collection processes (Ruíz et al., 2010). Around the world and in recent decades, the consumption of nickel–cadmium batteries has increased and because of it, pollution has also increased; mentioning an example: Mexico sells 600 million batteries per year and it is estimated that in the last four decades, an average of 20.69 tons of Cd were discarded to the environment.

In Europe, the use of hydrometallurgical and pyrometallurgical techniques is the main process to recycle the Cd in batteries; the pyrometallurgical techniques separate and transform it through thermal treatment in a reducing medium. This technique is simple although it spends high amounts of energy; on the other hand, the hydrometallurgical techniques use acids or bases for the total or partial dissolution of the metals. In Mexico, these techniques are not considered viable due to costs and the infrastructure and also because technology for recycling is not available (Castro and Díaz, 2004).

9.4 Removal Methods for Cd Present in Water and Soil

Most of the treatments that exist are for removing Cd in water, and the nature and physicochemical properties of the water bodies to be treated have to be considered; the most used treatments for the removal of heavy metals are: filtration, reverse osmosis, ion exchange, chemical precipitation, adsorption, electrodeposition, evaporation, flotation, and extraction of liquid; however, these treatments in many cases do not reach the necessary final result since the permissible concentrations of heavy metals in water are too low (NOM-002-ECOL-1996) in the case of Cd, the permissible index in water is <5 ppm (Peña, 2016; Ruíz et al., 2010).

In the case of Cd soil contamination, several technologies have been implemented to remove heavy metals; most of them are on-site processes because they eliminate the use of transport, excavation and use of reactors; the most used treatments are electroremediation, fluid drag, forced extraction with vapors, chemical oxidation, and landfarming. Electroremediation is the most used treatment because it can remove some organic pollutants and has a greater efficiency in the removal of inorganic contaminants in comparison to the other techniques.

In addition to being applicable to different types of soils, it should be noted that the percentage of removal will depend on the type of soil and the pollutant(s) present, but depending on the soil, regeneration may be required through the addition of fertilizers and organic matter (De la Rosa al., 2007; García et al., 2011). Among the most used treatments for the elimination of Cd in water and soil are the following:

Chemical precipitation: technique to eliminate Cd in water that uses mainly carbonate, iron hydroxide, or aluminum hydroxide, which react with the Cd forming agglomerates; there are some disadvantages of this technique, like when organic substances are present their performance decreases, it is not selective, and it is necessary to use coagulants and flocculants to carry out the separation (Peña, 2016).

Ionic exchange: treatment in water where there is a chemical reaction in which one ion is exchanged by another with similar load; the absorption force is determined by the hydrated radius, the ones with the lowest hydrated radius or those with the greatest possibility of water loss are more strongly retained. It should be considered that hydration is directly proportional to the ion charge and inversely proportional to the ionic radius. The disadvantage of this method is that when the concentration of the metal is not high, it will not be effective. Besides that, in the presence of calcium, sodium, and magnesium, the yield is lower because it also reacts with the resin saturating it, and in addition, the resins are not tolerant to pH changes and a previous treatment is necessary to eliminate suspended materials (Peña, 2016).

Adsorption: technique implemented in water; it can be by physical or chemical fixation. Normally, activated carbon and zeolites are used. This methodology offers good results and flexibility in the design and operation; however, depending on the material used, the costs can be high. It is necessary to eliminate suspended material so that a previous treatment is needed and the adsorption capacity of the material depends on the pH (Peña, 2016).

Electroremediation: a technique used in soils, useful to remove organic and inorganic contaminants through the use of an electric field in a matrix treated with water or an electrolyte; the current passes from an anode to a cathode and transports the pollutant in this way. The technique allows a directed migration and three phenomena occur, namely, electroosmosis, electromigration, and electrophoresis (Peña, 2016; Rebollo, 2012).

These technologies can become inefficient and expensive when dissolved metal concentrations are in the order of 1–100 mg/L, which is why it has been necessary to search for efficient biotechnologies, with lower costs and friendly to the environment.

9.5 Biotechnology for the Removal of Cd in Contaminated Environments

In recent years, various methodologies and materials that use organisms have been studied for their role in the removal and transformation of heavy metals from contaminated sites such as water bodies and soils, as an alternative to conventional methods, which in some cases have been shown to be deficient and expensive; so the implementation of organisms with metabolic capabilities skilled of transforming these compounds is important to counteract the harmful effects of these metals and their bioavailability in the environment. The organisms mostly used for metal removal include bacteria, fungi and plants; organisms use enzymatic and non-enzymatic mechanisms for metal removal their removal capacity being superior to that reported by conventional physicochemical techniques and sometimes very selective.

The use of biological agents to remove toxic pollutants from the environment is known as bioremediation and is one of the safest, most cost-effective, and environmental-friendly technologies and the most effective tools to treat contaminated soils. Organisms have different mechanisms to absorb or transform various pollutants because they can use these as a source of energy or nutrients; among the enzymatic mechanisms of microorganisms are oxidation, reduction, methylation, and demethylation. In the case of heavy metals, these cannot be degraded but transformed, changing their oxidation state; cadmium in oxidation state $+1$ is unstable, which reduces mobility or increases its solubility, facilitating its elimination (Sánchez and Martínez, 2016).

Among the microorganisms most studied and used for the processes of bioremediation of heavy metals are bacteria; an important group are the sulphate-reducing bacteria, which are used to treat water and soil contaminated with metals because in anaerobic conditions, they oxidize organic compounds when using sulfate as an electron acceptor and producing sulfur (S^{2-}), which reacts with the metals present and forms precipitates with lower toxicity like metal-sulfur. Filamentous fungi have shown a great ability to absorb metals in aqueous solutions where their capacity will depend on pH, biomass, and the metals present; *Rhizopus*, *Penicillium*, and *Phanerochaete* are among the most commonly used genera.

The effectiveness of the bioremediation processes is related to the characteristics of the contaminated site and the pollutants present; to treat effluents and industrial waste that have more than one contaminant, systems have been implemented that integrate different technologies, such as the use of fungi and bacteria in processes of phytoremediation, achieving favorable results since it increases and improves the removal rate (Covarrubias, 2017; Garzón et al., 2017). Summarizing some of the most commonly used bioremediation techniques are as follows:

Chelation: process used for different types of metals including cadmium; it is based on the formation of stable metal complexes with chelating agents synthesized by microorganisms or plants. In the case of plants, the root is the main entry of the metals that arrive by diffusion, massive flow, or cation exchange because the root presents a negative charge for the carboxyl groups, which interact with the positive charges of the metals creating a dynamic equilibrium and facilitating the entrance to the cells by means of apoplastic or simplistic way; once inside, they form complexes with ligands.

Some of the chelating compounds in plants are organic acids such as citric, oxalic, and malic acid, some amino acids such as histidine and cysteine and peptides such as phytochelatins and metallothionines (Delgadillo et al., 2011). On the other hand, microorganisms can synthesize biosurfactants which decrease the solubility and mobility of the metal. In the case of Cd, the rhamnolipids synthesized by *P. aeruginosa* are used, which have shown specificity for Cd (Sánchez and Martínez, 2016).

Biolixiviation: process based on the ability of microorganisms to increase the solubility of the metal, thus allowing its extraction from the medium. Once the metal has been leached, it can be recovered by conventional physical–chemical techniques such as ion exchange, precipitation, or biosorption; some fungi that can carry out this process for Cd are *Aspergillus*, *Fusarium*, *Curvalaria*, and *Penicillium* that, through organic acids such as oxalic acid, propionic acid, succinic acid, and acetic acid, allow the solubilization of metals (Sánchez and Martínez, 2016).

Phytoremediation: it is a method that uses plants to remove metals from soil, air, or water. It is based on biochemical processes of plants and the microorganisms associated with them to remove, reduce, transform, mineralize, degrade, volatilize, or stabilize contaminants (Delgadillo et al., 2011).

There are three main mechanisms by which plants can remove contaminants: physical (filtration, adsorption, volatilization, and sedimentation), chemical (precipitation, hydrolysis, oxidation–reduction, or photochemical reactions), and biological (by microbial metabolism, metabolism of the plant or bioabsorption); according to their mode of action, they can be classified as phytoextraction, which is the concentration of the metal in the leaves, fruits, and stems; rhizofiltration is the absorption of the metal by the root; phytostabilization, when the plant can reduce or prevent metal mobility; phytostimulation, when the plant through exudates favors the proliferation of microorganisms capable of transforming the metal; phytovolatilization, when the plant can concentrate and modify the metal to a less toxic state and then release it to the environment (Kulshreshtha et al., 2014; Sánchez and Martínez, 2016).

Most of the studies for Cd removal by plants are based on the ability of the plants to absorb the metal; there is a wide variety of plants reported with high capacity to accumulate metals known as hyperaccumulators; in the case of cadmium, they must be able to accumulate at least 100 µg/g of Cd to enter in this classification (Kulshreshtha et al., 2014). Table 9.1 lists some hyperaccumulating plants of some metals among which Cd has been a target metal.

One of the disadvantages of phytoremediation is that plants have little biomass, small size, and shallow roots, which has led to the use of genetically modified plants that achieve higher percentages of removal; some of the modified plants are *Nicotiana tabacum* with the *CUP1* and *TapCSI* genes, Brassica, the genes *gshI* and *gshII*, and Arabidopsis thaliana with the *SAT* and *ASTL* genes (Serrano et al., 2008).

Biosorption: it is based on the capture of metals by physicochemical interactions of the metal with cell surface ligand by adsorption processes or ion exchange due to functional groups; these may be weak as lipids or glucans because the interaction is by the group hydroxyl, or strong in the case of amino acids, proteins, and carboxylic acids. For Cd, studies have shown favorable results with the use of fungi, yeasts, algae, and plants. The algae, mostly composed by polysaccharides, proteins, and lipids, have many functional groups that can react with metal ions, for which reason they have been subject of study, being more efficient than activated carbon and zeolites (Areco, 2012; Sánchez and Martínez, 2016). Table 9.2 summarizes some of the studies conducted regarding the removal of Cd contained in wastewater by bioadsorption processes; this type of streams is the most difficult to treat because they contain more than one pollutant.

Table 9.1   Some plant species able to hyper accumulate metals

Species	Metals	Reference
Thlaspi caerulescens	Cd, Zn	Pirzadah et al. (2015)
Lemna minor	Cd, Zn, Cu, Pb	Arenas et al. (2011)
Pistias tratiotes	Cd, Cr, Cu, Hg	Delgadillo et al. (2011)
Brassica juncea	Cd, Cr, Cu, Hg	Bradkariya et al. (2014)
Athyrium yokoscense	Cd, Cu, Pb, Zn	Chen et al. (2009)
Noccaea caerulescens	Zn, Cd	Martos et al. (2016)
Noccaea brachypetala	Zn, Cd	Martos et al. (2016)
Echornia crassipes	Pb, Cu, Cd, Fe	Zaranyika and Ndapwadza (1995)
Vallisneria americana	Cr, Cd, Pb	Singh and Tripathi (2007)
Bacopa monnier	Cr, Cd, Cu, Hg, Pb	Sinha and Chandra (1990)
Haumaniastrum robertii	Cr, Cd, Hg, Pb, Zn	Singh and Tripathi (2007)
Festuca arundinacea	Cd, Cr, Co, Cu, Ni, Pb, Zn	Arenas et al. (2011)
Artemisiiofolia ambrosia	Pb, Cd, Cr, Cu	Singh and Tripathi (2007)
Salix viminalis	Pb, Cd, Cr, Co, Cu, Ni, Pb, Zn	Hammer et al. (2003)
Salvinia molesta	Zn, Cd, Cr, Co, Cu	Singh and Tripathi (2007)
Silene vulgaris	Zn, Cd, Cr, Cu, Pb	Pradas de Real et al. (2011)
Spirodela polyrhiza	Zn, Cd, Cr, Co, Ni, Pb	Singh and Tripathi (2007)
Caerulescens thlaspi	Zn, Cd, Cr, Co, Cu, Pb	Delgadillo et al. (2011)
Trifolium pratense	Zn, Cd, Cr, Co, Cu, Ni, Pb	Singh and Tripathi (2007)

Table 9.2   Removal studies of Cd in industrial wastewaters

Absorbent	Removal (%) or (mg/g)	Characteristics	Reference
Wallnut shell (*Carya illinuensis*)	72.35%	Higher Cd (II) removal at pH 6, 150 min	Kamar and Nechifor (2015)
Sea lettuce (*Ulva lactana*)	25 mg/g algae	Higher absorption of Cd using algae as a bioabsorbent (100 mg Cd/10 mg algae), 3 h	Areco (2012)
Banana hull (*Musa cavendishii*)	70%	Residues of banana hull were oxidized with sulfuric acid and used as adsorbent of 100 mg Cd at pH 4, 25 min	Peña (2016)
Algae (*Ascophyllum nodosum*)	529.4 mg/g algae	High percentages were obtained at pH 4.9, 30 min	Navarro et al. (2004)
Orange peel (*Citrus sinensis*)	98%	High percentage at pH 4–6, 25°C using 100 mg Cd/L (0.89 mM)	Pérez et al. (2007)

Currently, Cd is recognized as one of the heavy metals that causes major problems to the environment and human health; however, the efforts made to control Cd emissions to the environment are focused on the metal recycling or confinement because the treatments implemented for the removal of Cd both in soil and in water can be expensive and do not meet the desired concentrations, so the implementation of biotechnology for Cd removal is an option to achieve high removal percentages, with lower costs and with less ecological impact. It should be noted that the type of methodology used will depend on the type of contaminants present, the concentration of contaminants, and the nature of the contamination site.

9.6 Use of Biotechnology for the Treatment of Cd in Residues

Residues or waste containing Cd are mainly of two types: solids and Cd suspended in liquid. Within the solid wastes, there can be found batteries and mine tailings primarily, while the residues suspended in liquids can be found in industrial wastewaters, livestock sewage, agricultural runoff, and mining wastewaters.

Most of the solid wastes are confined or recycled; mentioning the case of batteries, developed countries have plants that, through hydrometallurgical or pyrometallurgical techniques, recover the metals contained in them, including the Cd which can recover up to a purity of 99.9% and is reused to manufacture new batteries, and countries that do not have the infrastructure and technology for their treatment are limited to confining the batteries or sending them to countries that can carry out the process. Mining wastes mainly agglomerate in the form of sludge and they are deposited in tailings dams; these residues are difficult to treat because they contain a large number of metals and have the disadvantage that there is the migration of contaminants to soil or water and most of them are incinerated. Other methodologies for the disposal of mining wastes are the disposal of dry or thickened tailings in deposits, landfills of underground mines, open pits, and sub-aqueous disposition (Caviedes et al., 2015; Hernandez et al., 2009). The use of biotechnology to treat solid waste with Cd is scarcely implemented mainly because the volumes generated are large and contain different heavy metals and other types of pollutants.

Currently, some industries have decided for biotechnological processes to treat their wastewater through various treatments, the most common being bioabsorption with different organisms, from the use of dead or living microbial biomass to plants. Several studies report that Gram-negative bacteria can accumulate Cd ions as is the case of *P. aeruginosa* and *Alcaligenes faecalis*; however, Gram-positive bacteria such as *Bacillus megaterium*, *Bacillus pumilus*, *Bacillus subtilis*, and *Brevibacterium iodinium* have shown to be able to absorb considerable amounts of Cd as well as some fungi such as the case of *Coprinopsis atramentaria* (Igiri et al., 2018).

Bioabsorption with plants is another methodology widely used in water contaminated with Cd; the use of algae has been proposed mainly because they have shown removal percentages up to 90% as well as the implementation of artificial wetlands containing algae. The advantages of these methodologies are the low operating cost; metals can be recovered, high percentages of removal and versatility.

References

Areco, M.M., 2012. Biosorcion, una opcion eficiente y sustentable para la remediación ambiental, Producción y Ambiente, 7, 580–584.

Arenas, A.D., Marcó, L.M., Torrres, G., 2011. Evaluación de la planta Lemna minor como biorremediadora de aguas contaminadas con mercurio. Avances en Ciencias e Ingeniería 2, 1–11.

CAMIMEX, 2012. Cadmio Cd. Revista oficial de la Cámara minera de México [WWWDocument]. URL:https://www.camimex.org.mx/index.php/secciones1/sala-de-prensa/uso- de-los-metales/cadmio/

CAMIMEX, 2016. Participación de México en la producción minera mundial en 2106. Revista oficial de la Cámara minera de México [WWWDocument]. URL: https://www.camimex.org.mx/files/2615/0092/9348/05-Info17.pdf

Carballo, M.E., Martínez, A., Salgado, I., Pérez, L., Cruz, M., Liva, M.B., Alleyne, S., Rodríguez, M.M., Garza, Y., 2017. Standardization of variables involved in cadmium and zinc microbial removal from aqueous solutions. Biotecnol. Apl. 34, 1221–1225.

Castro, J.C., Diaz, M.L., 2004. La contaminación por pilas y baterias en México. Gaceta Ecológica, 72, 53–74.

Caviedes, D.I., Muñoz, R.A., Gualtero, A.P., Rodriguez, D., Sandoval, J., 2015. Tratamientos para la Remoción de Metales Pesados Comúnmente Presentes en Aguas Residuales Industriales. Una Revisión. Ing. Región, 13(1), 73–90.

Chen, Z., Setagawa, M., Kang, Y., Sakurai, K., Aikawa, Y., Iwasaki, K., 2009. Zinc and cadmium uptake from a metalliferous soil by a mixed culture of *Athyrium yokoscense* and *Arabis flagellosa*. Soil Sci. Plant Nutr. 55, 315–324. https://doi.org/10.1111/j.1747-0765.2008.00351.x

Covarrubias, S., 2017. Contaminación ambiental por metales pesados en México: problemática y estrategias de fitorremediación. Rev. Int. Contam. Ambient. 33, 7–21. https://doi.org/10.20937/RICA.2017.33. esp01.01

De La Rosa, D.A., Teutli-León, M.M.M., Ramírez-Islas, M.E., 2007. Electrorremediación de suelos contaminados, una revisión técnica para su aplicación en campo. Rev. Int. Contam. Ambient. 23, 129–138.

Delgadillo, A.E., González, C.A., Prieto, F., Villagómez, J.R., Acevedo, O., 2011. Fitorremediación: una alternativa para eliminar la contaminación. Trop. Subtrop. Agroecosystems 14, 597–612.

Frías, E., Osuna, L., Voltolina, D., López, L., Izaguirre, F., 2009. Cadmium, copper, lead and zinc content of the mangrove oyster *Crassostrea cortenziensis* of seven coastal lagoons of NW México. Bull Environm. Contam. Toxicol, 83, 593–599.

Frías, M.G., Osuna, J.I., Aguilar, M., Voltolina, D., 2010. Cadmio y plomo en organismos de importancia comercial de la zona costera de Sinaloa, México: 20 años de estudios. CICIMAR Oceánides 25, 101–110.

García, L., Vargas Ramírez, M., Reyes Cruz, V., 2011. Electrorremediación de suelos arenosos contaminados por Pb, Cd y As provenientes de residuos mineros, utilizando agua y acido acético como electrolitos. Superf. Vacío 24, 24–29.

Garzón, J.M., Rodríguez, J.P., Hernández, C., 2017. Aporte de la biorremediación para solucionar problemas de contaminación y su relación con el desarrollo sostenible 19, 309–318. https://doi.org/10.22267/rus.17 1902.93

Godoy, E., 2011. México esconde el cadmio debajo de la alfombra [WWW Document]. IPS Agencia Not. URL http://www.ipsnoticias.net/2011/11/ mexico-esconde-el-cadmio-debajo-de-la-alfombra/ (accessed 4.8.19).

Hammer, D., Kayser, A., Keller, C., 2003. Phytoextraction of Cd and Zn with *Salix viminalis* in field trials. Soil Use Manag. 19, 187–192. https://doi.org/10.1111/j.1475-2743.2003.tb00303.x

Hernández, A., Mondragón, R., Cristóbal, A., Rubias P., Robledo, S., 2009. Vegetación, residuos de mina y elementos potencialmente tóxicos de un jal de Pachuca, Hidalgo, México. Revista Chapingo Serie Ciencias Forestales y del Ambiente. 15(2), 109–114.

Igiri, B., Okoduwa, S., Idoko, G., Akabuogo, E., Adeyi, A., Ejiogu, K., 2018. Bioremediation of heavy metals contaminated ecosystem from Tannery waste water: A review. J. Toxicol., 2018, 1–16. https://doi.org/10.1155/2018/2568038.

Kamar, F.H., Nechifor, A.C., 2015. Removal of copper ions from industrial wastewater using walnut shells as a natural adsorbent material U.P.B. Sci. Bull., 77(3), 141–150.

Kulshreshtha, A., Agrawal, R., Barar, M., Saxena, S., 2014. A review on bioremediation of heavy metals in contaminated water. J. Environ. Sci., Toxicol. Food Technol. 8(7), 44–50.

Martelli, A., Rousselet, E., Dycke, C., Bouron, A., Moulis, J.M., 2006. Cadmium toxicity in animal cells by interference with essential metals. Biochimie 88, 1807–1814. https://doi.org/10.1016/j.biochi.2006.05.013

Martos, S., Gallego, P., Sáez, H., López, A., Cabot, C., Puschenrieder, C., 2016. Characterization of Zinc and Cadmium hyperaccumulation in three *Noccaea* (Brassicaceae) populations from non-metalliferous sites in the eastern Pyrenees. Front. Plant Sci. 7, 128. doi: 10.3389/fpls.2016.00128

Navarro, A., Blanco, D., Llanos, B., Flores, J., Maldonado, H., 2004. Biorremoción de cadmio (II) por desechos de algas marinas: optimización del equilibrio y propuesta de mecanismo. Soc. Quím. Perú 70, 147–157.

Norma Oficial Mexicana NOM-002-ECOL-1996. Que establece los limites maximos permisibles de contaminantes para las aguas residuales tratadas que se reusen en servicios al público. 1996. 15.

Peña, D., 2016. Remoción de Cd por *Musa cavendishii*, L. Thesis, Universidad Autónoma del Estado de México.

Pérez, A.B., Zapata, V.M., Ortuño, J.F., Aguilar, M., Sáez, J., Lloréns, M., 2007. Removal of cadmium from aqueous solutions by adsorption onto orange waste. J. Hazard. Mater. 139, 122–131. https://doi.org/10.1016/j.jhazmat.2006.06.008

Pirzadah, T.B., Malik, B., Tahir, I., Kumar, M., Varma, A., Rehman, R.U., 2015. Phytoremediation, in: Soil Remediation and Plants. Elsevier, pp. 107–129. https://doi.org/10.1016/B978-0-12-799937-1.00005-X

Pradas de Real, A., Alarcón, R., García, P., Gil, D., Lobo, M., Pérez, S., 2011. Cd and Cr uptake by six genotypes of *Silene vulgaris* (Moench) Garcke. Proceedings of 15th ICHMET, Phytoremediation, 563–566.

Rebollo, J., 2012. Eliminación de Cadmio (II) de efluentes urbanos tratados mediante procesos de bioadsorción: el efecto competitivo de otros metales pesados. Master Thesis, Universidad Politécnica de Cartagena, Colombia.

Ruiz-López, V., González-Sandoval, M.R., Barrera-Godínez, J.A., Moeller-Chávez, G., Ramírez-Camperos, E., Durán-Domínguez-de-Bazúa, M.C., 2010. Cadmium and zinc removal from a mining reprocessing aqueous stream using artificial wetlands. Tecnología, Ciencia y Educación 25(1): 27–34.

Sánchez, G., Martínez, M.Á.C., 2016. Thesis. Ecotoxicología del cadmio; riesgo para la salud de la utilización de suelos ricos en cadmio.

Serrano, R., Martínez de la Casa, Romero, P., Del Río L., Sandalio, L., 2008. Toxicidad del Cadmio en Plantas. Ecosistemas. 17(13), 139–146.

Singh, S.N., Tripathi, R.D., 2007. Environmental bioremediation technologies. Springer Science & Business Media.

Sinha, S., Chandra, P., 1990. Removal of Cu and Cd from water by Bacopa monnieri L. Water Air Soil Pollut. 52(3), 271–276.

Wright, J., 2003. Environmental chemistry. New York, NY. Routledge.

Zaranyika, M.F., Ndapwadza, T., 1995. Uptake of Ni, Zn, Fe, Co, Cr, Pb, Cu and Cd in water hyacinth (*Eichhornia crassipes*) in mukuvisi and manyame rivers, Zimbabwe. J. Environ. Sci. Health Part A Environ. Sci. Eng. Toxicol. 30, 157–169. https://doi.org/10.1080/10934529509376193

Index

About the Editor

Norma Gabriela Rojas Avelizapa holds a degree in Industrial Chemistry from the Universidad Veracruzana (1995), a Master's degree in Biotechnology from CINVESTAV (1995), and a PhD in Environmental Biotechnology from CINVESTAV (1999). From 1999 to 2015 she worked as a Scientific Researcher at the Mexican Petroleum Institute in the area of petroleum biotechnology. She belongs to the National Research System (SNI) with level II. Since 2005, she has been working as a Titular Professor at the National Polytechnic Institute in the area of Environmental Biotechnology. Her research interests include bioremediation of contaminated sites, valorization of industrial wastes, bioleaching of industrial wastes, production of metallic and organic nanoparticles by biological means. She has written more than 50 papers in national and international journals and published book chapters. She has developed 20 research projects, participated in the edition of 2 technical books, 2 patent applications, 2 technological developments in collaboration with the Mexican Petroleum Institute.